破冰利器·双语精装版

神奇的
答案之书

焦海利 ◎ 著

中国致公出版社·北京

图书在版编目（CIP）数据

神奇的答案之书 / 焦海利著 . -- 北京：中国致公出版社，2024.1（2024.10 重印）

ISBN 978-7-5145-2186-3

Ⅰ.①神… Ⅱ.①焦… Ⅲ.①心理压力-心理调节-通俗读物 Ⅳ.① B842.6-49

中国国家版本馆 CIP 数据核字 (2023) 第 213351 号

神奇的答案之书 / 焦海利著
SHENQI DE DAAN ZHI SHU

出　　版	中国致公出版社
	（北京市朝阳区八里庄西里 100 号住邦 2000 大厦 1 号楼西区 21 层）
发　　行	中国致公出版社（010-66121708）
责任编辑	胡梦怡
责任校对	魏志军
装帧设计	天下书装
责任印制	宋洪博
印　　刷	晟德（天津）印刷有限公司
版　　次	2024 年 1 月第 1 版
印　　次	2024 年 10 月第 2 次印刷
开　　本	880 毫米 × 1230 毫米　1/32
印　　张	20
字　　数	20 千字
书　　号	978-7-5145-2186-3
定　　价	49.80 元

（版权所有，违者必究，举报电话：010-82259658）
（如发现印装质量问题，请寄本公司调换，电话：010-82259658）

使用指南

1. 将《神奇的答案之书》合上,用双手夹住。

2. 想好你要解决的问题,然后闭上双眼,将这个问题在心中默默重复。

3. 约 10 秒之后,用右手大拇指和食指将书翻到任意一页。

4. 睁开双眼,呈现在你眼前的就是你要的答案!

5. 重复上述步骤,寻找其他问题的答案。

想好问题了吗?

请合上本书,

开启你的神奇答案探索之旅吧!

▽

这具有重要意义

It is of great significance

▲

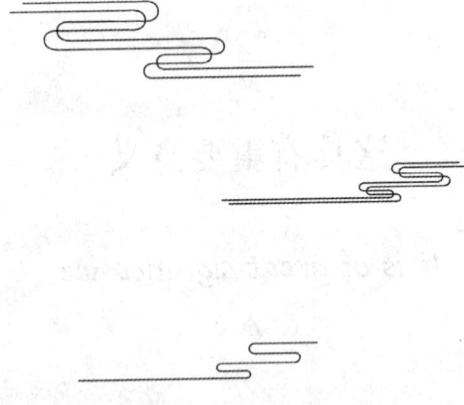

神奇的答案之书
THE BOOK OF MAGIC ANSWERS

神奇的答案之书
THE BOOK OF MAGIC ANSWERS

▽

防止意外发生

Prevent accidents from occurring

▲

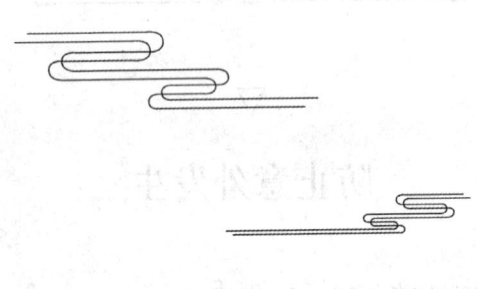
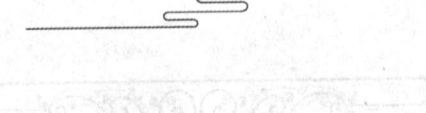

一切将取决于你的选择

It will all depends on your choice

▽

转移注意力

Divert your attention

▲

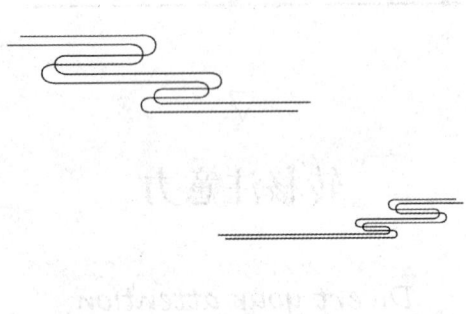

神奇的答案之书
THE BOOK OF MAGIC ANSWERS

▽

离开

Leave

▲

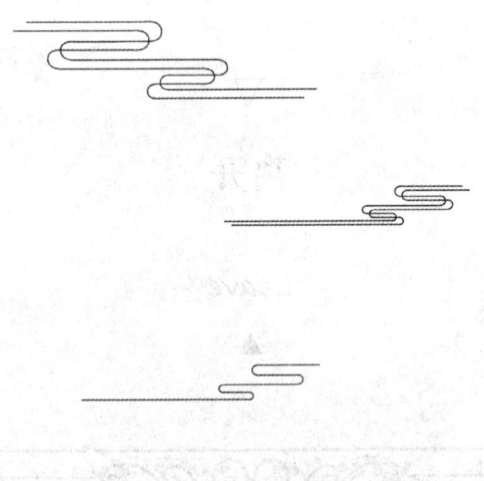

神奇的答案之书
THE BOOK OF MAGIC ANSWERS

神奇的答案之书
THE BOOK OF MAGIC ANSWERS

▽

你需要其他人的帮助

You need help from others

▲

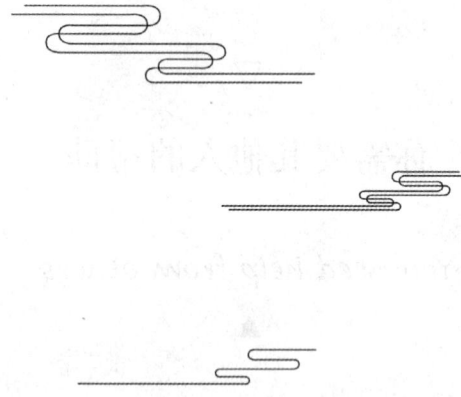

神奇的答案之书
THE BOOK OF MAGIC ANSWERS

▽

这有些特别

This is peculiar

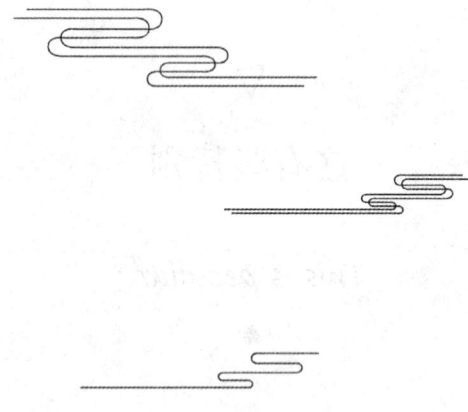

神奇的答案之书
THE BOOK OF MAGIC ANSWERS

过段时间就不那么重要了

After a time, it will not
be so important

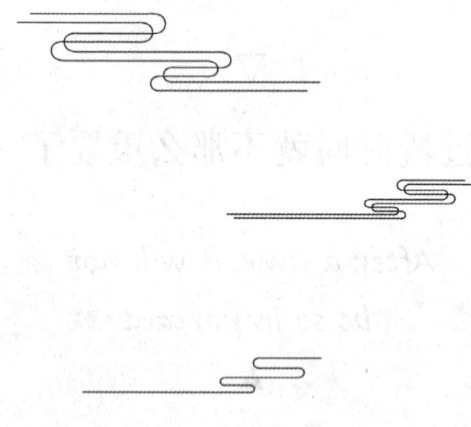

神奇的答案之书
THE BOOK OF MAGIC ANSWERS

▽

不要犹豫

Don't hesitate

▲

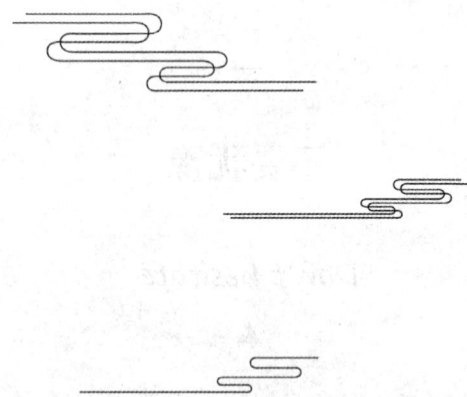

神奇的答案之书
THE BOOK OF MAGIC ANSWERS

▽

先完成其他事

Get other things done first

▲

▽

给自己一点时间

Give yourself some time

▲

神奇的答案之书
THE BOOK OF MAGIC ANSWERS

▽

现在你就能

You are able to do it now

▲

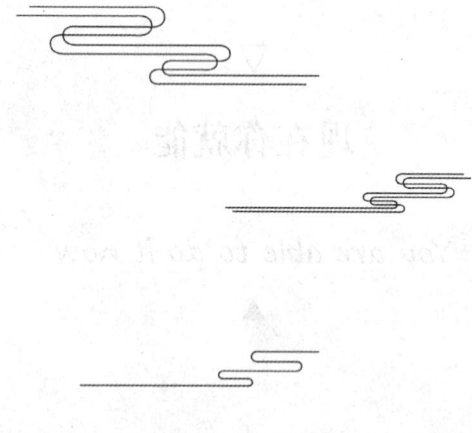

▽

那不值得心烦

That is not worth
your vexation

▲

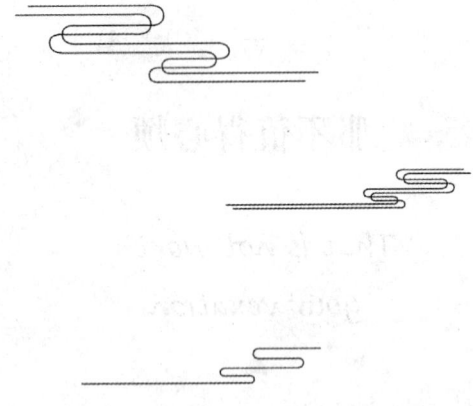

▽

那将影响别人对你的看法

That will affect others'
views on you

▲

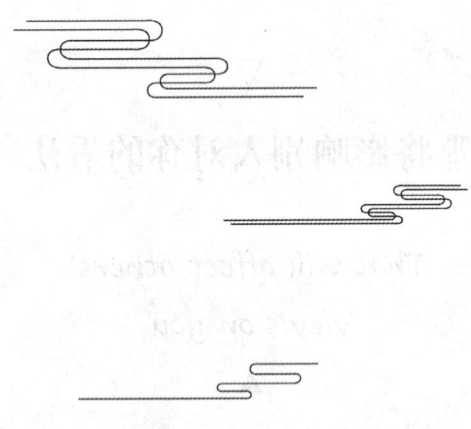

神奇的答案之书
THE BOOK OF MAGIC ANSWERS

▽

照别人说的去做

Do what others have told you

▲

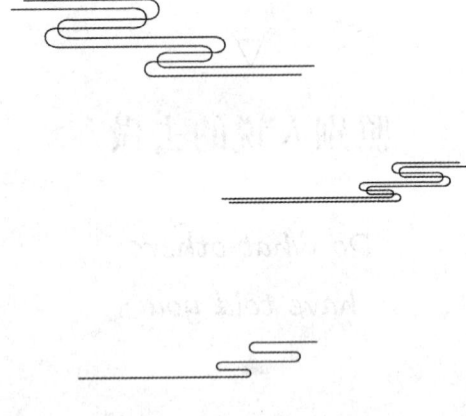

▽

转移你的注意力

Divert your attention

▲

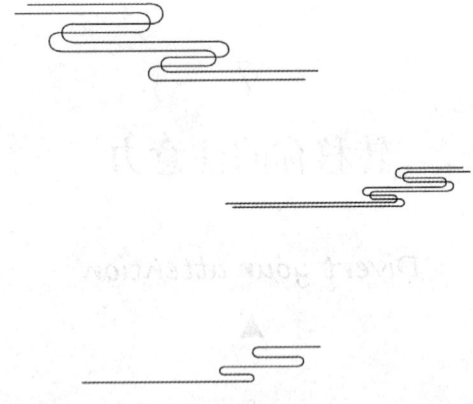

▽

你会失望的

You will be disappointed

▲

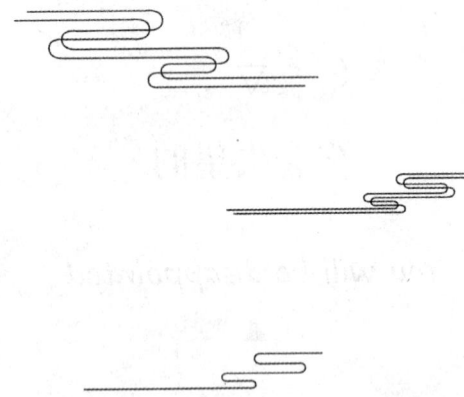

神奇的答案之书
THE BOOK OF MAGIC ANSWERS

最好关注你的工作

You'd better focus your attention on your work

神奇的答案之书
THE BOOK OF MAGIC ANSWERS

▽

形势尚不明朗

The situation is still not very clear

▲

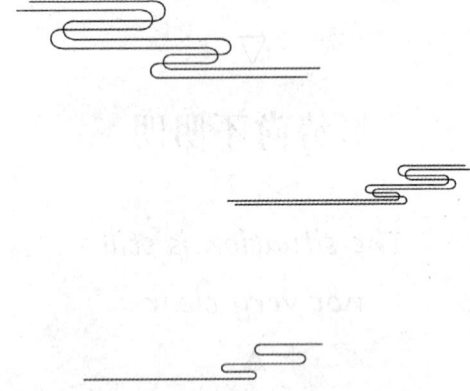

神奇的答案之书
THE BOOK OF MAGIC ANSWERS

神奇的答案之书
THE BOOK OF MAGIC ANSWERS

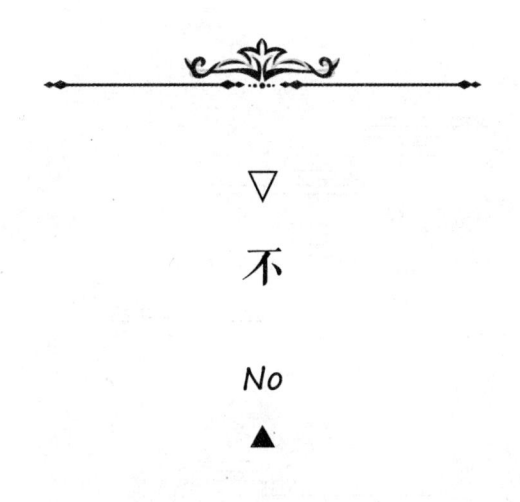

▽

不

No

▲

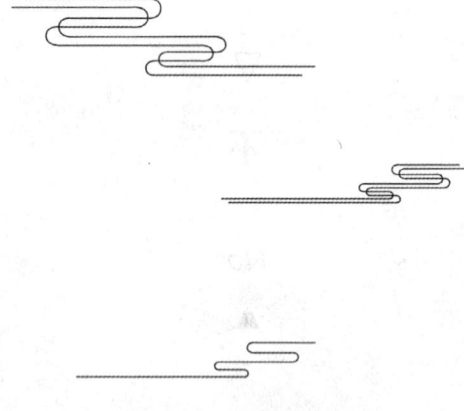

神奇的答案之书
THE BOOK OF MAGIC ANSWERS

▽

结果可能令人吃惊

The result might
be surprising

▲

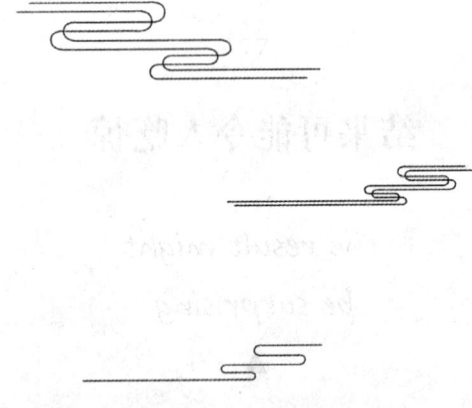

神奇的答案之书
THE BOOK OF MAGIC ANSWERS

▽

有决心就能成功

Having determination,
anyone can succeed

▲

▽

你的行动会使一切得到改善

Your action will make
everything get well

▲

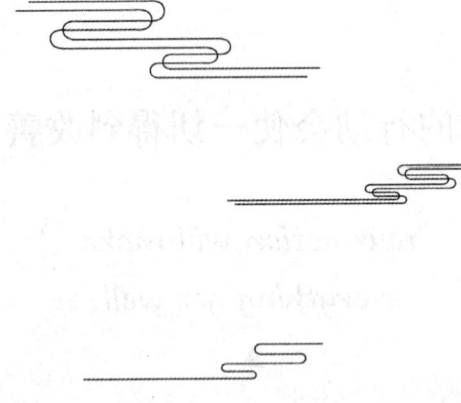

神奇的答案之书
THE BOOK OF MAGIC ANSWERS

▽

做一次改变

Make a change

▲

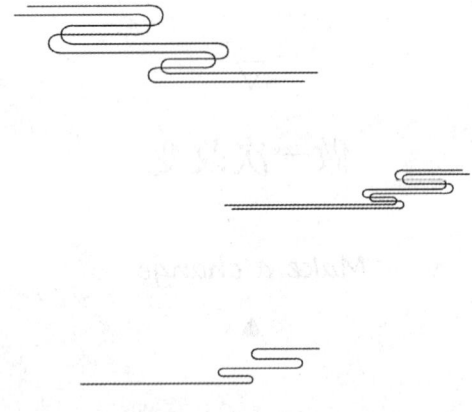

照别人告诉你的去做

Do what others have told you

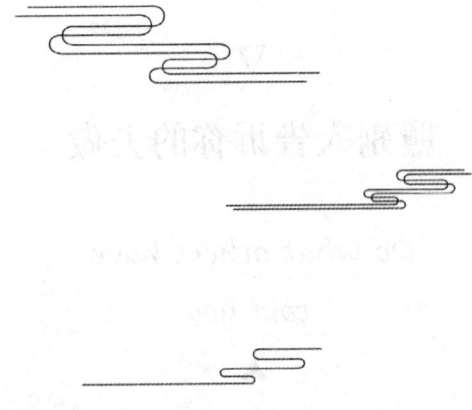

▽

不能保证

*It can not
be guaranteed*

▲

神奇的答案之书
THE BOOK OF MAGIC ANSWERS

答案可能会以
另一种形式出现

The answer may be
given in another form

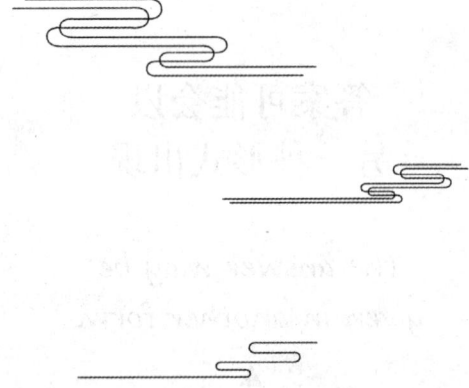

神奇的答案之书
THE BOOK OF MAGIC ANSWERS

▽

毋庸置疑

Don't doubt it

▲

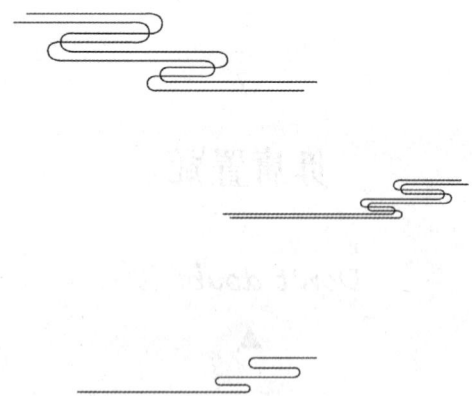

神奇的答案之书
THE BOOK OF MAGIC ANSWERS

这样做会使事情变得有趣

That will make things interesting

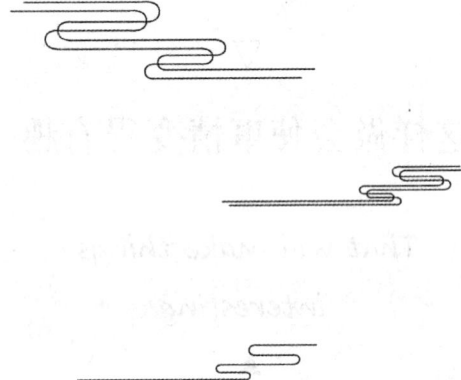

神奇的答案之书
THE BOOK OF MAGIC ANSWERS

▽

这是肯定的

It's sure

▲

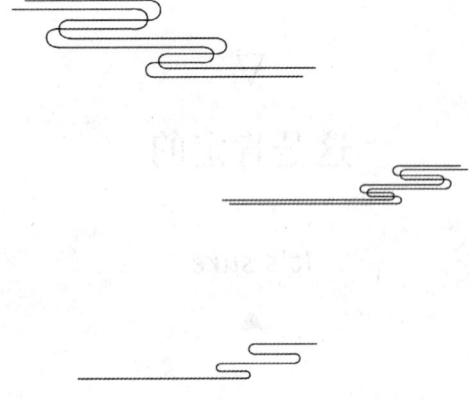

神奇的答案之书
THE BOOK OF MAGIC ANSWERS

有可能会伤害到他人

It might do harm to others

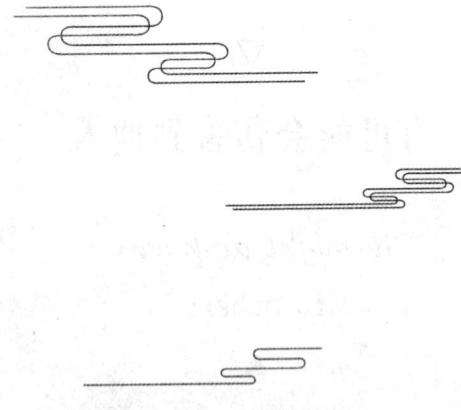

神奇的答案之书
THE BOOK OF MAGIC ANSWERS

那只会浪费钱

That will only
waste money

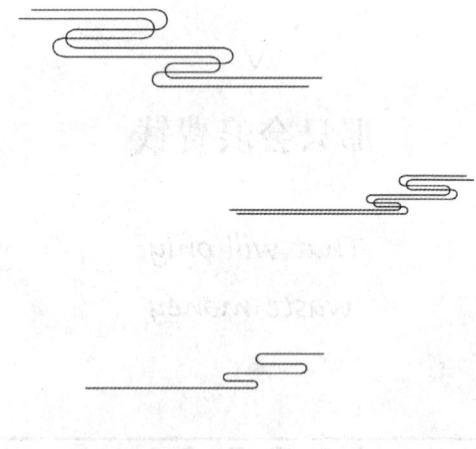

神奇的答案之书
THE BOOK OF MAGIC ANSWERS

▽

最好等待

You'd better wait

▲

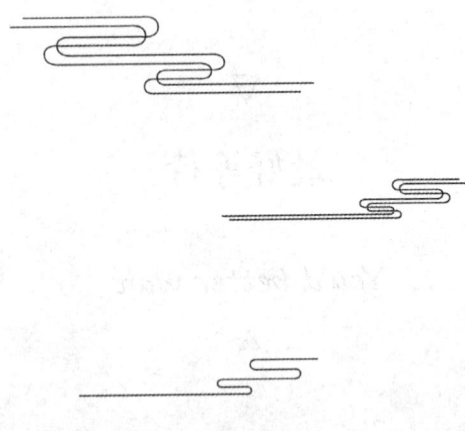

神奇的答案之书
THE BOOK OF MAGIC ANSWERS

$\triangledown$

本页前第三个答案

The third answer before this page

▲

神奇的答案之书
THE BOOK OF MAGIC ANSWERS

神奇的答案之书
THE BOOK OF MAGIC ANSWERS

▽

那超出了你的控制

This is beyond your
control

▲

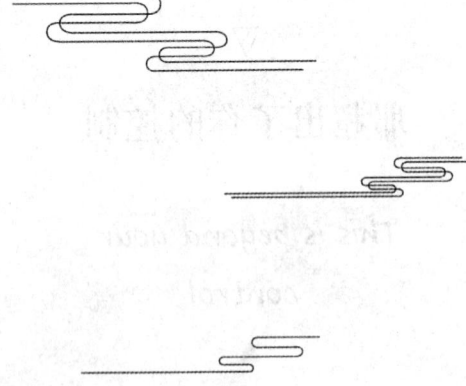

神奇的答案之书
THE BOOK OF MAGIC ANSWERS

你需要采取主动

You need to take the initiative

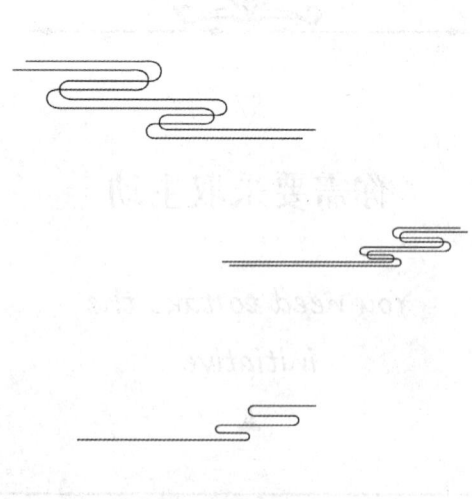

▽

似乎没问题

It seems all right

▲

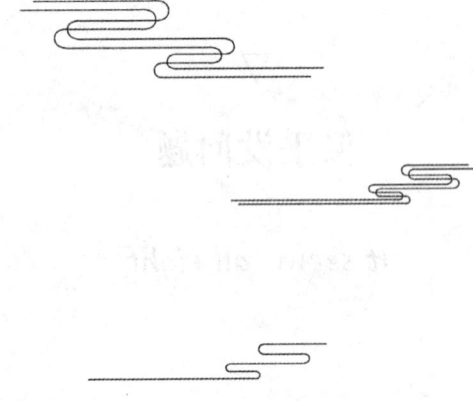

▽

肯定

It's sure

▲

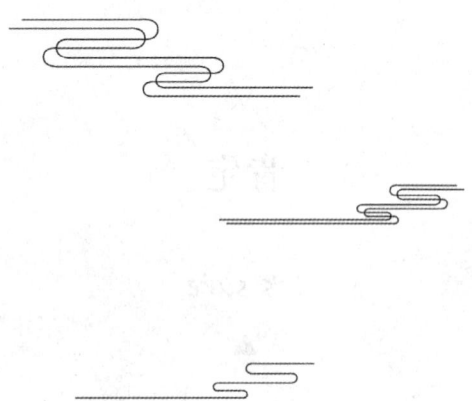

▽

不要在意

Don't care about it

▲

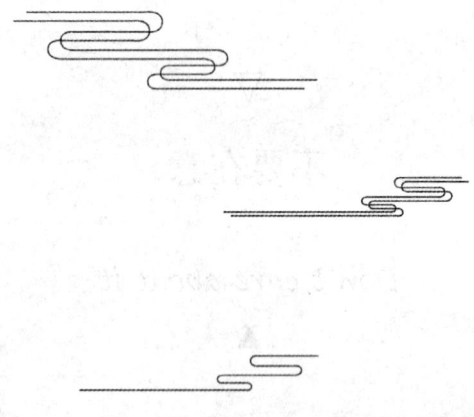

神奇的答案之书
THE BOOK OF MAGIC ANSWERS

尽早做好它

Get it done as early
as possible

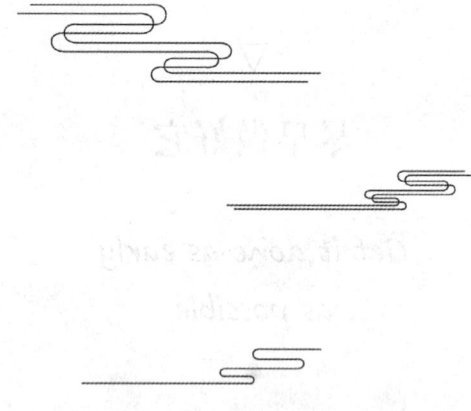

你终会发现你想知道的一切

Finally you will find out what you want to know

神奇的答案之书
THE BOOK OF MAGIC ANSWERS

▽

除非你独自一人做

Unless you do it alone

▲

神奇的答案之书
THE BOOK OF MAGIC ANSWERS

▽

是，但不要强求

Yes, but don't force it

▲

▽

你需要去适应

You need to adapt
yourself to it

▲

神奇的答案之书
THE BOOK OF MAGIC ANSWERS

▽

明天再试试

Try again tomorrow

▲

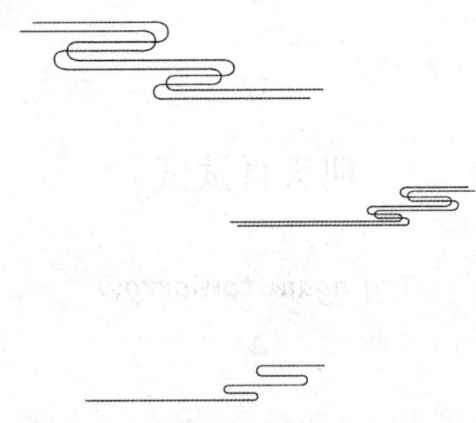

以更放松的态度去面对

Face it with a more relaxed attitude

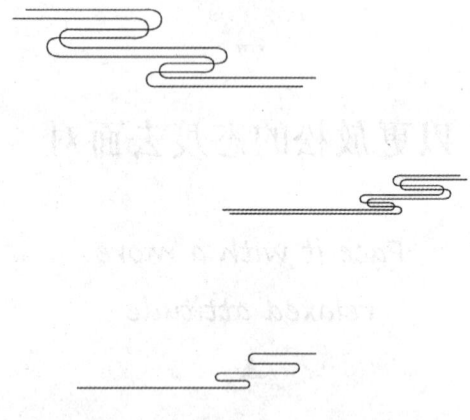

▽

你需要考虑其他方法

You need to consider
other methods

▲

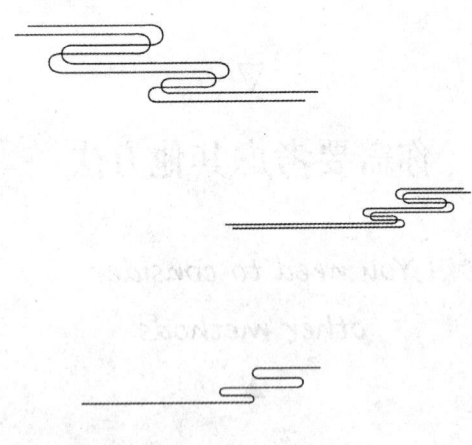

神奇的答案之书
THE BOOK OF MAGIC ANSWERS

▽

习惯于接受改变

Be accustomed to accepting change

▲

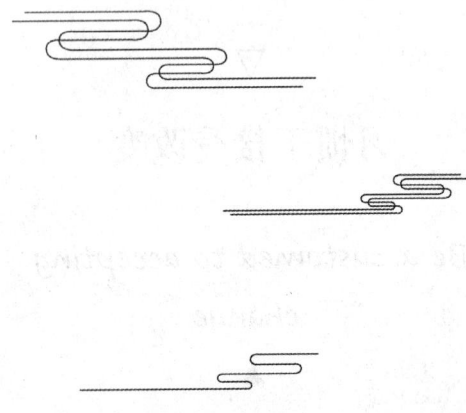

▽

别浪费时间了

Don't waste time anymore

▲

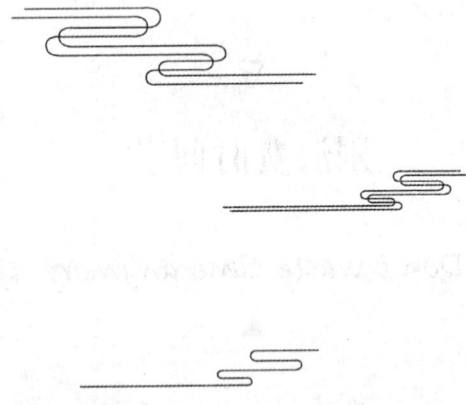

神奇的答案之书
THE BOOK OF MAGIC ANSWERS

▽

是的

Yes

▲

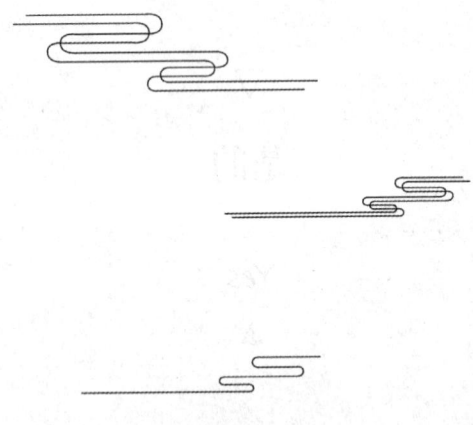

神奇的答案之书
THE BOOK OF MAGIC ANSWERS

本页前第一个答案

The first answer before
this page

寻求更多的选择

Seek more choice

▲

神奇的答案之书
THE BOOK OF MAGIC ANSWERS

这还不确定

That's not settled

神奇的答案之书
THE BOOK OF MAGIC ANSWERS

神奇的答案之书
THE BOOK OF MAGIC ANSWERS

▲

▽

谨慎处理

Treat it with caution

▲

▽

全力以赴

Try your best

▲

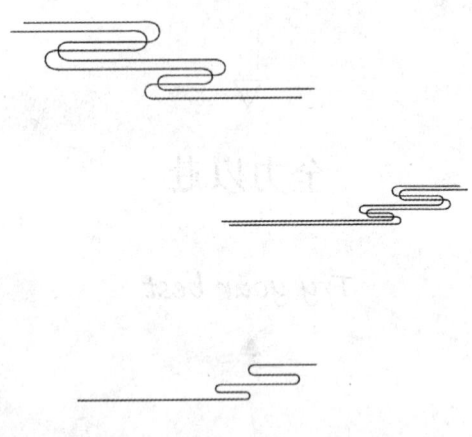

神奇的答案之书
THE BOOK OF MAGIC ANSWERS

给自己一点时间

Give yourself some time

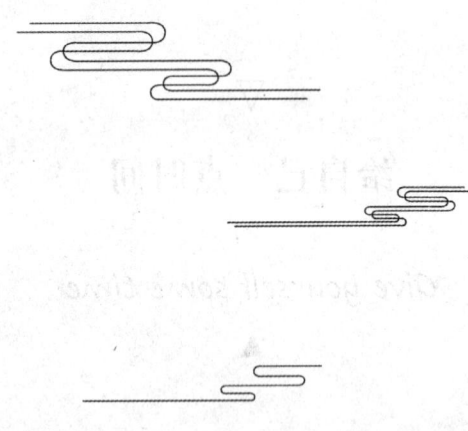

神奇的答案之书
THE BOOK OF MAGIC ANSWERS

神奇的答案之书
THE BOOK OF MAGIC ANSWERS

重新考虑你的做法

Reconsider what you did

神奇的答案之书
THE BOOK OF MAGIC ANSWERS

▽

问问你的母亲吧

Ask your mother about it

▲

问问你的母亲吧

Ask your mother about it

神奇的答案之书
THE BOOK OF MAGIC ANSWERS

不要独自一人去做

Don't do it by yourself

▽

或许当你长大些就知道了

Perhaps you will understand
it when you grow up a little

神奇的答案之书
THE BOOK OF MAGIC ANSWERS

▽

不要担忧

Don't worry

▲

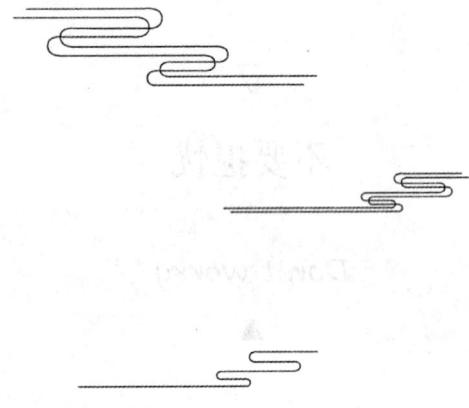

▽

保持开放的心态

Keep an open mind

▲

你会为自己所做的感到高兴的

You will be glad for what you have done

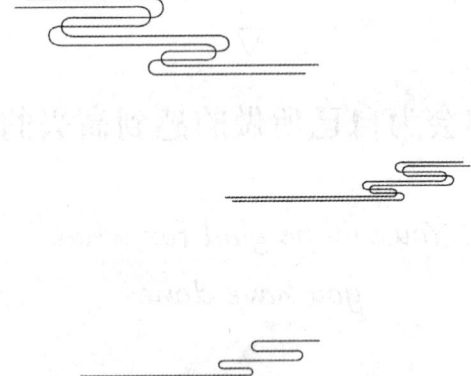

神奇的答案之书
THE BOOK OF MAGIC ANSWERS

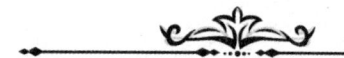

尽可能顺从你的意愿

Fall in with your wishes
as much as possible

问问你的母亲吧

Ask your mother about it

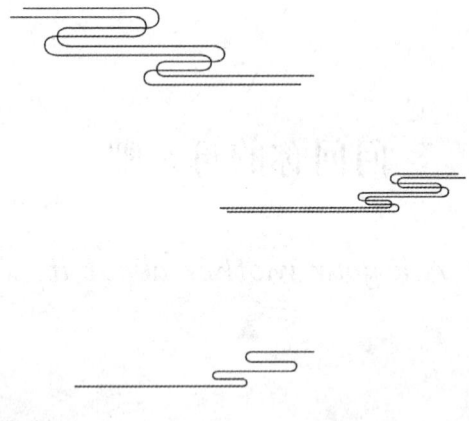

神奇的答案之书
THE BOOK OF MAGIC ANSWERS

▽

先做好自己应做的事

Do what you should do first

▲

神奇的答案之书
THE BOOK OF MAGIC ANSWERS

▽

不要怀疑

Don't doubt it

▲

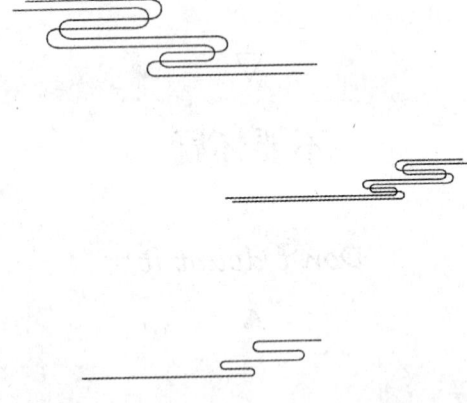

神奇的答案之书
THE BOOK OF MAGIC ANSWERS

神奇的答案之书
THE BOOK OF MAGIC ANSWERS

▽

是时候重新打算了

It's time to make
new plans

▲

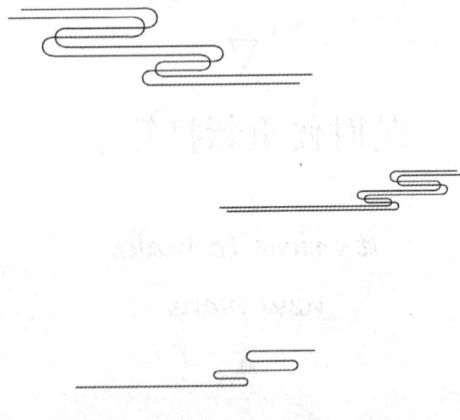

神奇的答案之书
THE BOOK OF MAGIC ANSWERS

▽

绝不

No way

▲

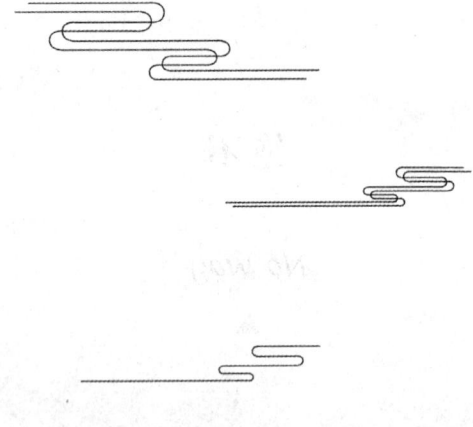

不要忽略显而易见的东西

Don't ignore those obvious things

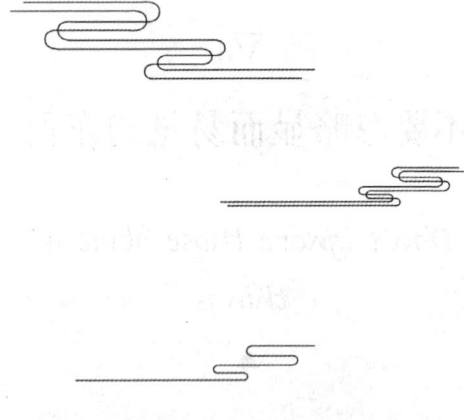

神奇的答案之书
THE BOOK OF MAGIC ANSWERS

省省力气吧

Don't waste your energy on it

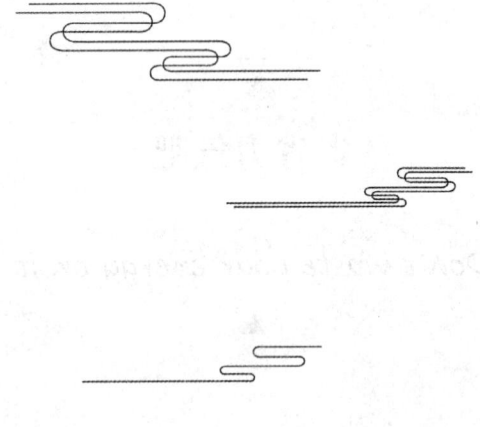

合作是关键

The key is cooperation

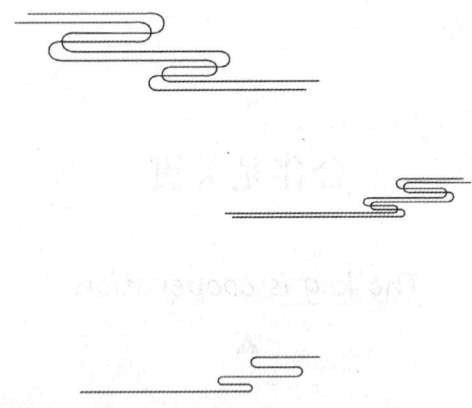

神奇的答案之书
THE BOOK OF MAGIC ANSWERS

▽

此时不宜

It's not suitable at this time

▲

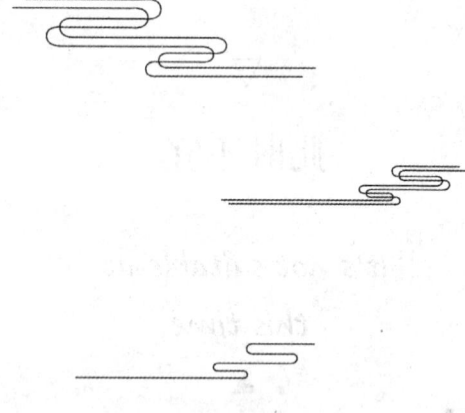

神奇的答案之书
THE BOOK OF MAGIC ANSWERS

▽

莫等待

Don't wait

▲

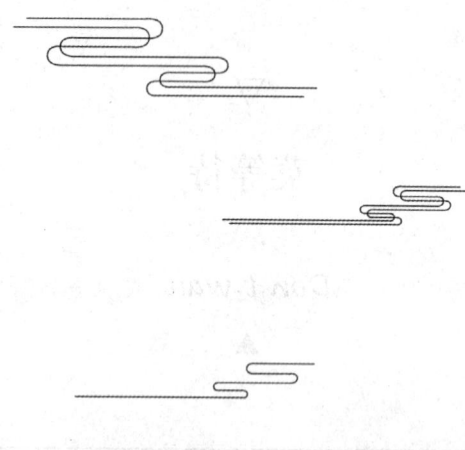

神奇的答案之书
THE BOOK OF MAGIC ANSWERS

神奇的答案之书
THE BOOK OF MAGIC ANSWERS

▽

灵活应对

Deal with it flexibly

▲

似乎已成事实

It seems to have become a fact

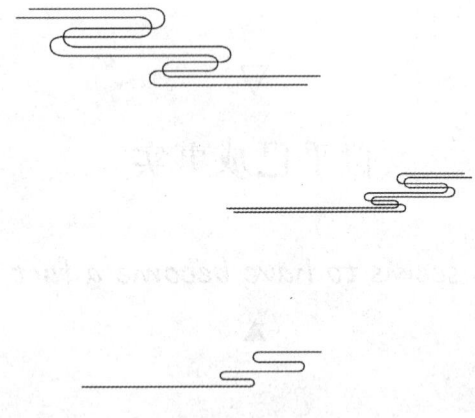

▽

轻松一点

Lighten up a little

▲

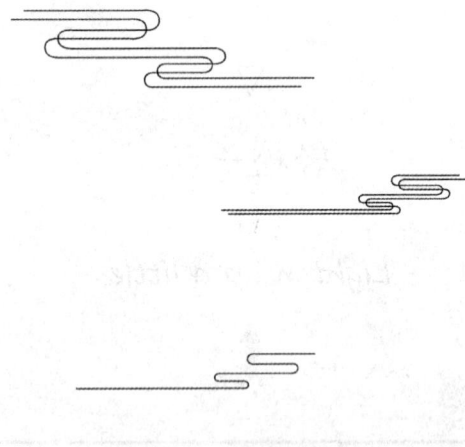

神奇的答案之书
THE BOOK OF MAGIC ANSWERS

避免第一个解决办法

Avoid the first solution

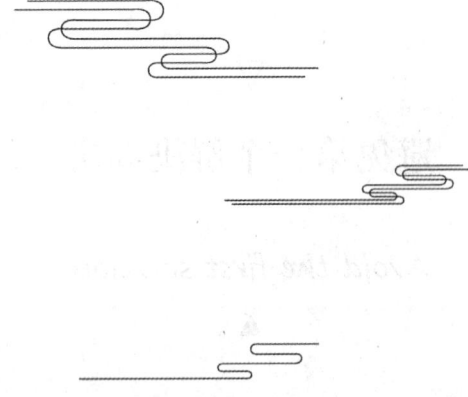

神奇的答案之书
THE BOOK OF MAGIC ANSWERS

神奇的答案之书
THE BOOK OF MAGIC ANSWERS

▽

随它去吧

Forget about it

▲

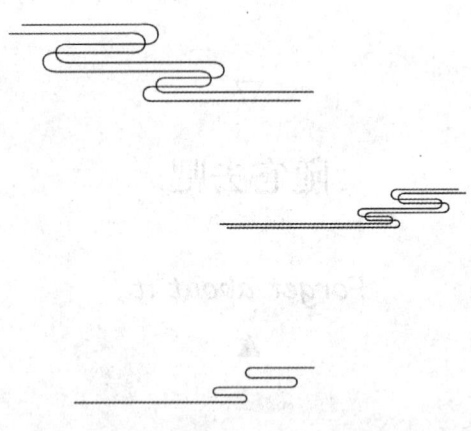

神奇的答案之书
THE BOOK OF MAGIC ANSWERS

▽

不要抱成见

Don't have any prejudice

▲

《莫慌乱》

Don't have any prejudice

神奇的答案之书
THE BOOK OF MAGIC ANSWERS

▽

你必须现在就行动

You must take action now

▲

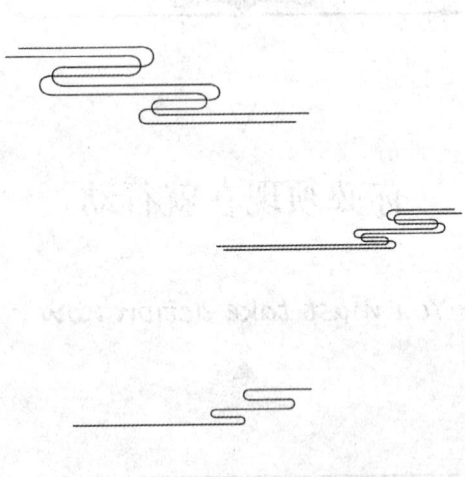

神奇的答案之书
THE BOOK OF MAGIC ANSWERS

▽

那可能很难，但值得

It may be difficult,
but it's worth it

▽

付出就会有回报

Your effort will be rewarded

▲

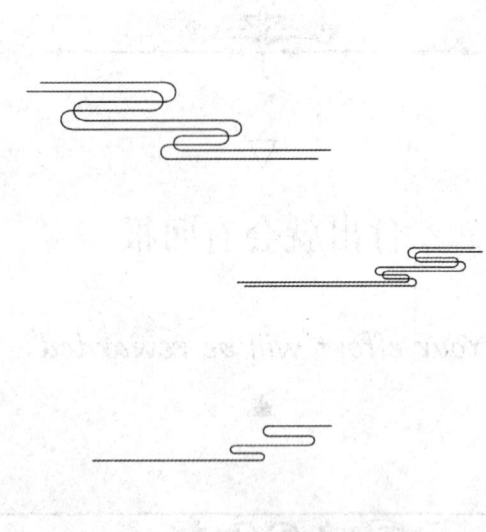

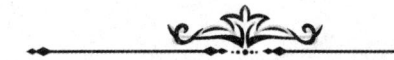

对意外要有思想准备

Be ready for the
unexpected

神奇的答案之书
THE BOOK OF MAGIC ANSWERS

神奇的答案之书
THE BOOK OF MAGIC ANSWERS

▽

更细心地去倾听，
你就会知道

If you listen to it more
carefully, you will know

▲

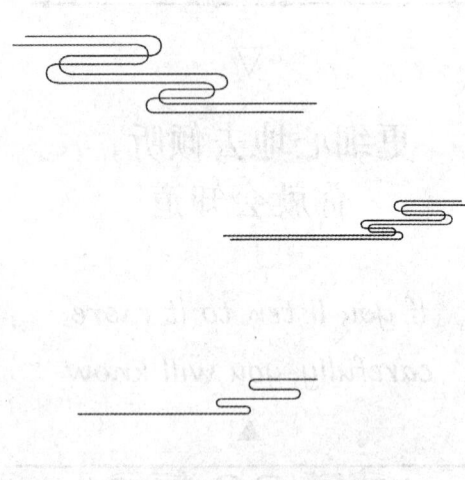

▽

且行且思

Think while you act

▲

神奇的答案之书
THE BOOK OF MAGIC ANSWERS

神奇的答案之书
THE BOOK OF MAGIC ANSWERS

▽

你有能力以任何方式改善

You are capable of improving
in every way

▲

神奇的答案之书
THE BOOK OF MAGIC ANSWERS

神奇的答案之书
THE BOOK OF MAGIC ANSWERS

▽

这些是你不会忘记的事物

These are things you would
not forget

▲

履行你的义务

Perform your duty

神奇的答案之书
THE BOOK OF MAGIC ANSWERS

快刀斩乱麻

Cut the Gordian Knot

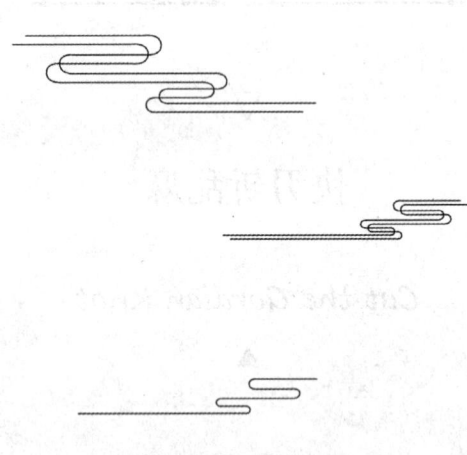

▽

别在这上面下赌注

Don't make a bet on this

▲

▽

也许吧

Perhaps

▲

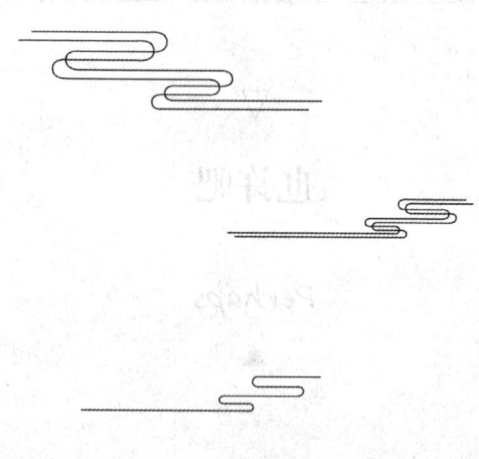

神奇的答案之书
THE BOOK OF MAGIC ANSWERS

▽

专注于你的家庭生活

Focus on your family life

▲

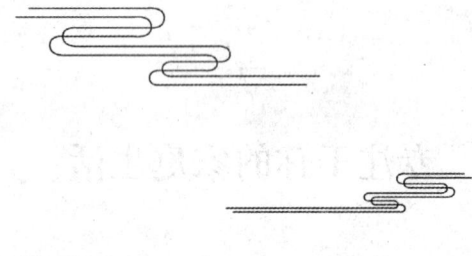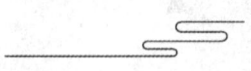

▽

绝对不行

Absolutely not

▲

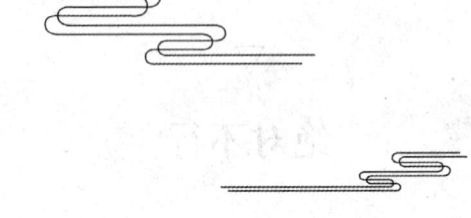
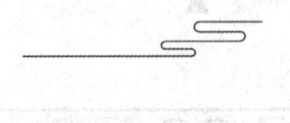

神奇的答案之书
THE BOOK OF MAGIC ANSWERS

神奇的答案之书
THE BOOK OF MAGIC ANSWERS

▽

等待

Wait

▲

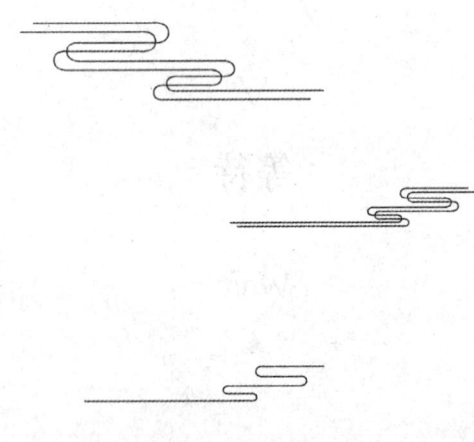

神奇的答案之书
THE BOOK OF MAGIC ANSWERS

▽

别犯傻了

Don't be silly anymore

▲

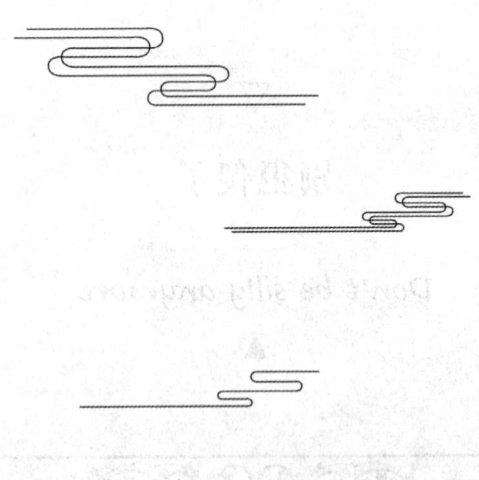

▽

你需要了解更多

You need to know more

▲

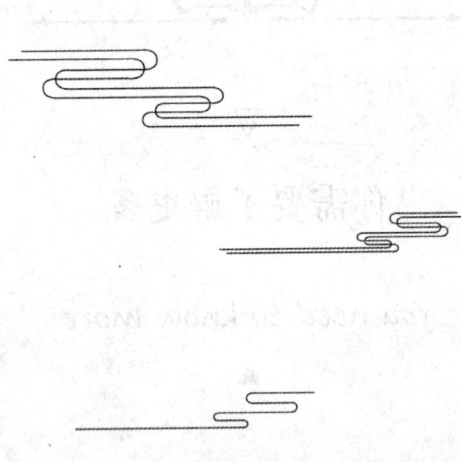

神奇的答案之书
THE BOOK OF MAGIC ANSWERS

▽

可能吧

Perhaps

▲

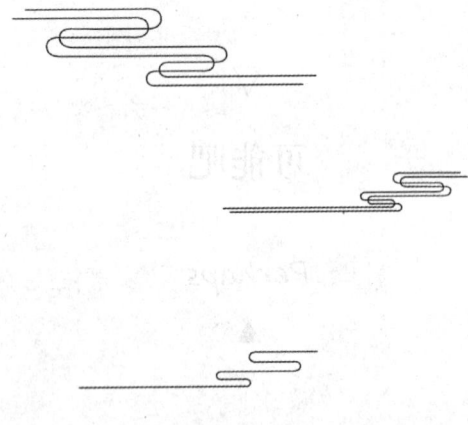

神奇的答案之书
THE BOOK OF MAGIC ANSWERS

表示怀疑

Be doubtful

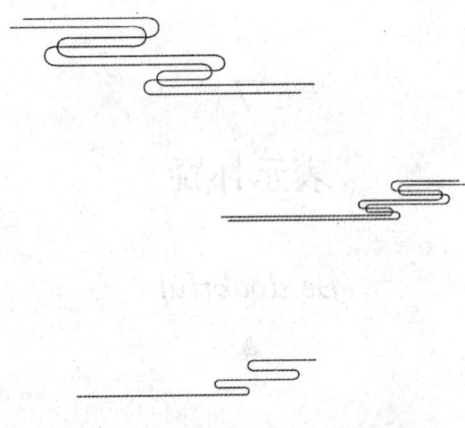

意义非凡

It's very important

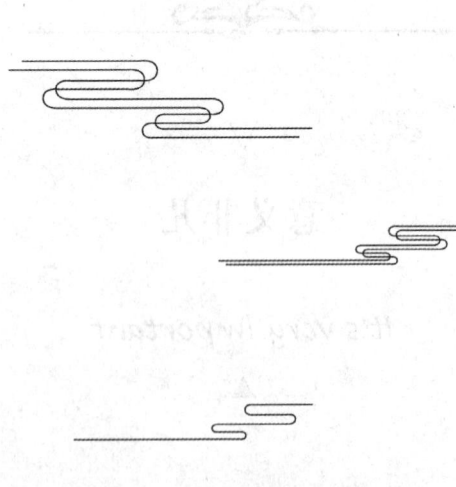

那可能非比寻常

It may be
extraordinary

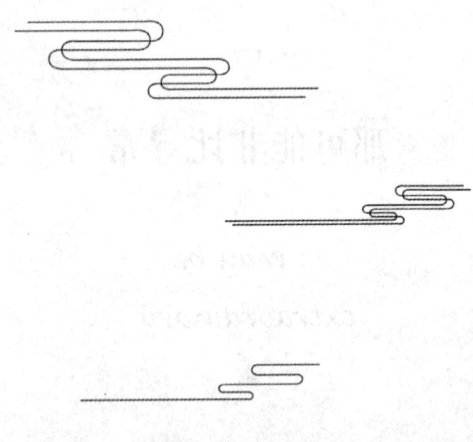

神奇的答案之书
THE BOOK OF MAGIC ANSWERS

不可能失败

It is not possible
to fail

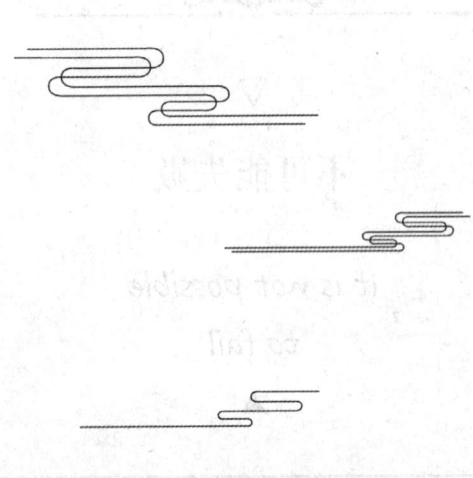

神奇的答案之书
THE BOOK OF MAGIC ANSWERS

神奇的答案之书
THE BOOK OF MAGIC ANSWERS

▽

情况很快会有变化

The situation will change soon

▲

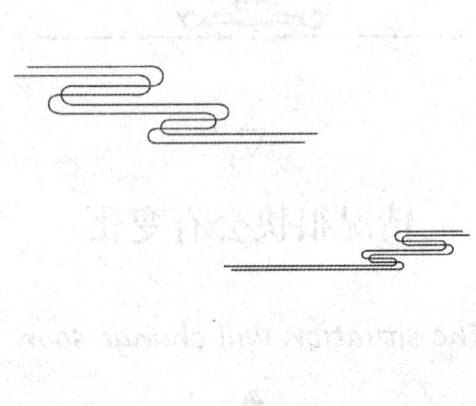

情况很快会发生变化

The situation will change soon

本页后第二个答案

The second answer after this page

这会带来好运

This will bring good luck

神奇的答案之书
THE BOOK OF MAGIC ANSWERS

神奇的答案之书
THE BOOK OF MAGIC ANSWERS

▽

顺其自然

Take it as it is

▲

▽

这时不要再自找麻烦

At this moment, don't get yourself into trouble

此时此刻,当自省察觉

At this moment, slow it get yourself into trouble.

神奇的答案之书
THE BOOK OF MAGIC ANSWERS

神奇的答案之书
THE BOOK OF MAGIC ANSWERS

▽

尝试一种更可能的
解决方案

Try a more reliable way
to solve it

▲

先做好其他事

Get other things done first

神奇的答案之书
THE BOOK OF MAGIC ANSWERS

▽

不要陷入消极情绪之中

Don't fall into the negative emotion

▲

○ ● ○ ● ● ● ○ ○ ● ○ ○ ● ● ●

请不要陷入消极情绪之中

Don't fall into the negative emotion

神奇的答案之书
THE BOOK OF MAGIC ANSWERS

柳暗花明又一村

Every cloud has a silver lining

▲

Every cloud has a silver lining.

神奇的答案之书
THE BOOK OF MAGIC ANSWERS

▽

你不得不妥协

You have to compromise

▲

列出否定的理由

List the negative reasons

▽

要有耐心

Have patience

▲

耐住性子

Have patience

神奇的答案之书
THE BOOK OF MAGIC ANSWERS

寻求更多选择

Seek more options

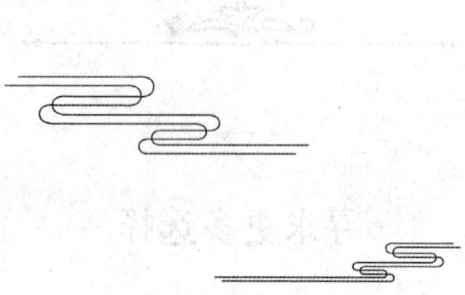
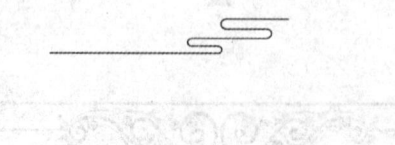

神奇的答案之书
THE BOOK OF MAGIC ANSWERS

▽

一笑置之

Carry off with a laugh

▲

▽

继续

Go on

▲

神奇的答案之书
THE BOOK OF MAGIC ANSWERS

别忘记享受乐趣

Don't forget to have fun

▲

神奇的答案之书
THE BOOK OF MAGIC ANSWERS

那是在浪费金钱

That's a waste of money

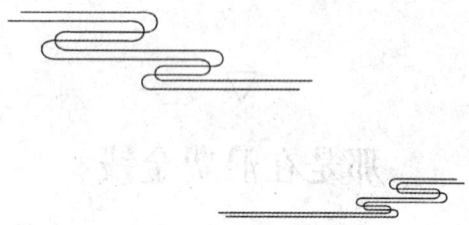
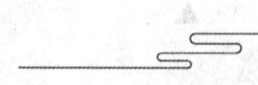

神奇的答案之书
THE BOOK OF MAGIC ANSWERS

▽

为了做出最好的决定，
务必保持冷静

In order to make the best
decision, you must keep calm

▲

神奇的答案之书
THE BOOK OF MAGIC ANSWERS

▽

尝试一个更没把握的方法

Try a less certain approach

▲

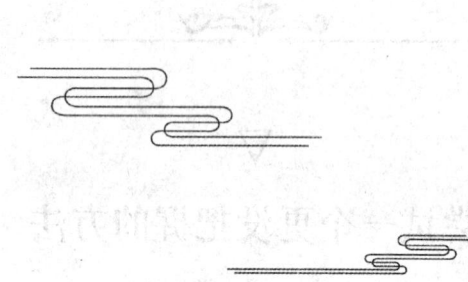

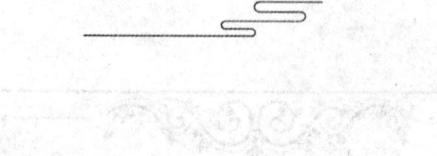

神奇的答案之书
THE BOOK OF MAGIC ANSWERS

清楚你自身的短板

Know your own weakness

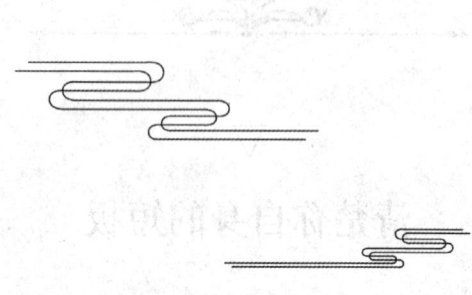

▽

那可能已成事实

It seems to have become a fact

▲

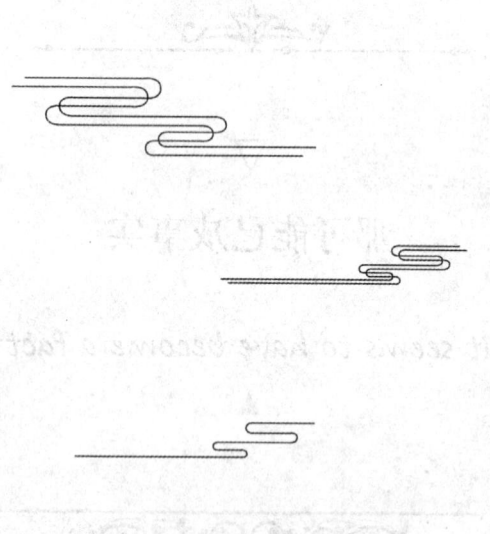

神奇的答案之书
THE BOOK OF MAGIC ANSWERS

▽

最好等待

You'd better wait

▲

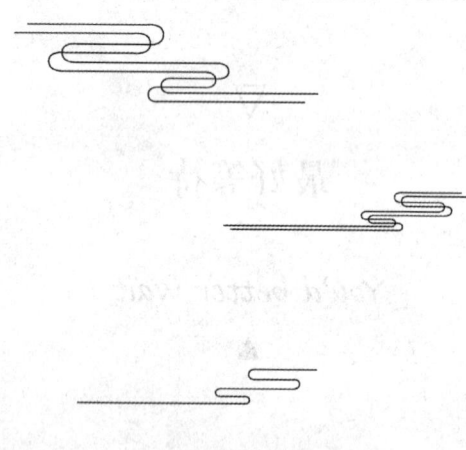

神奇的答案之书
THE BOOK OF MAGIC ANSWERS

神奇的答案之书
THE BOOK OF MAGIC ANSWERS

▽

先做重要的事

Important things have priority

▲

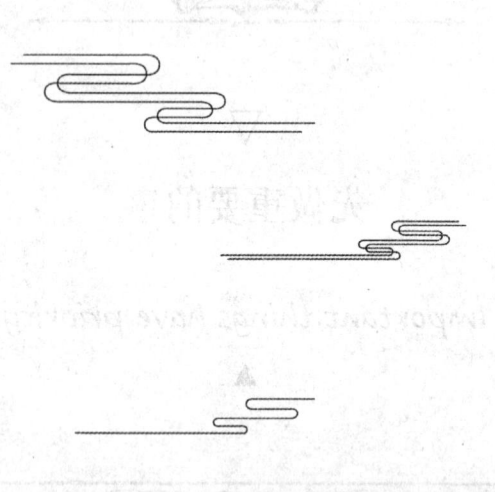

投硬币来做决定吧

Throw a coin to decide

▽

这不可能失败

It is not possible to fail

▲

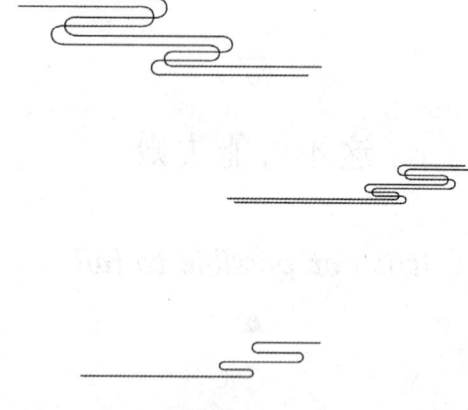

神奇的答案之书
THE BOOK OF MAGIC ANSWERS

▽

是，但不要强求不可能的事

Yes, but don't strive for the impossible

▲

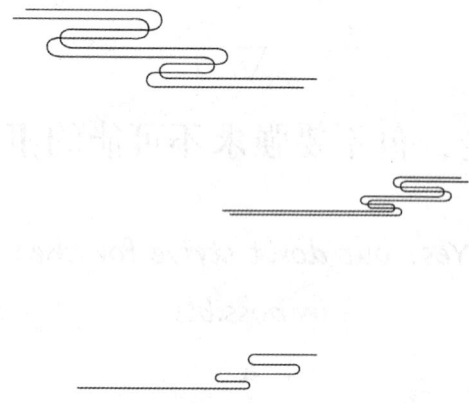

神奇的答案之书
THE BOOK OF MAGIC ANSWERS

▽

你最终能如愿

Eventually your wish may come true

▲

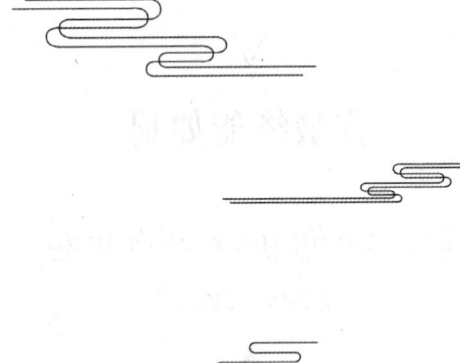

神奇的答案之书
THE BOOK OF MAGIC ANSWERS

神奇的答案之书
THE BOOK OF MAGIC ANSWERS

▽

可行

Practicable

▲

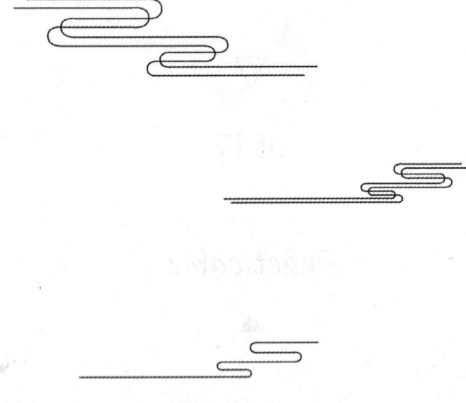

神奇的答案之书
THE BOOK OF MAGIC ANSWERS

▽

这时不要再自找麻烦

At this moment, don't get yourself into trouble

▲

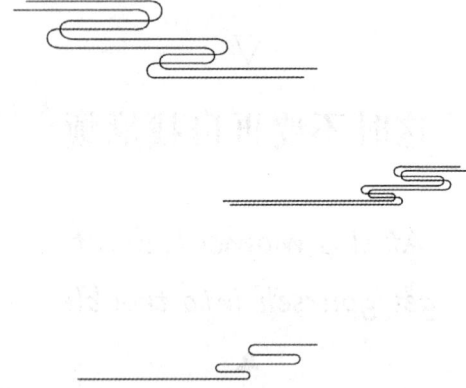

神奇的答案之书
THE BOOK OF MAGIC ANSWERS

现在你比以往任何时候
都看得更清楚

Now you see it more clearly
than ever before

神奇的答案之书
THE BOOK OF MAGIC ANSWERS

只须说声"谢谢"

You only need to say "thanks"

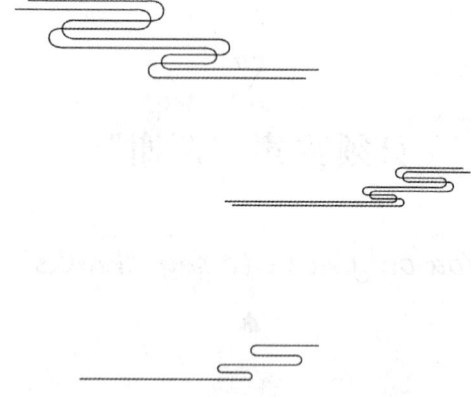

神奇的答案之书
THE BOOK OF MAGIC ANSWERS

或许，等你再年长些就明白了

Perhaps, you could understand it when you grow older

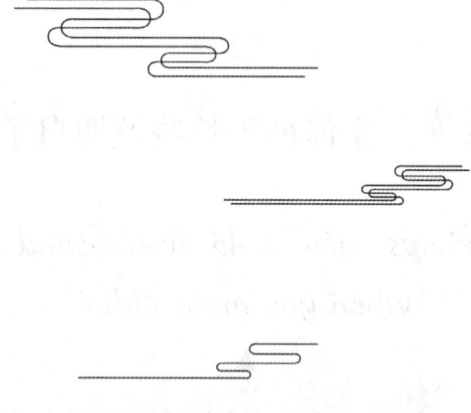

神奇的答案之书
THE BOOK OF MAGIC ANSWERS

这将轰动一时

That will cause a sensation

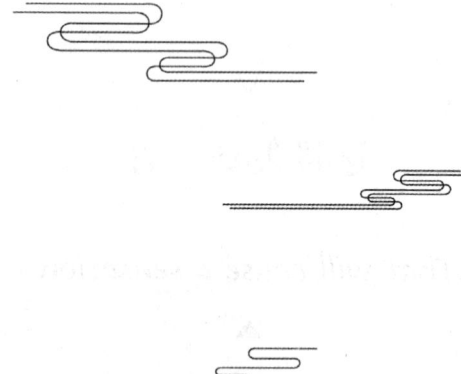

神奇的答案之书
THE BOOK OF MAGIC ANSWERS

▽

放手一搏

Go for it

▲

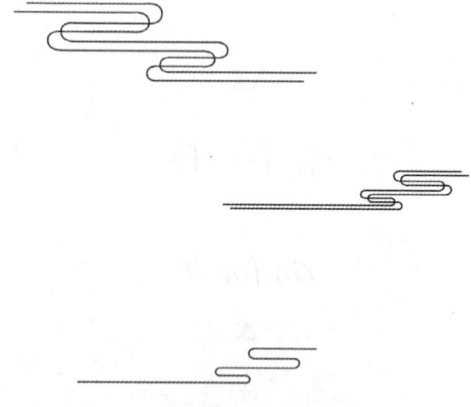

神奇的答案之书
THE BOOK OF MAGIC ANSWERS

神奇的答案之书
THE BOOK OF MAGIC ANSWERS

▽

事情会朝目标发展

Things will go your way

▲

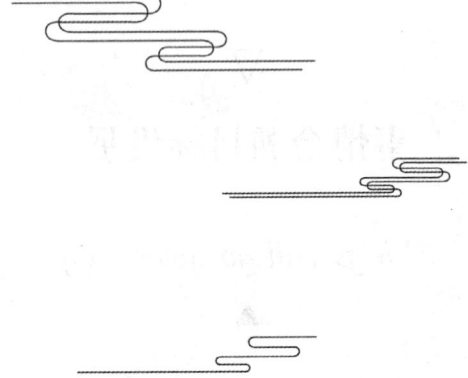

神奇的答案之书
THE BOOK OF MAGIC ANSWERS

▽

答案可能会以另一种形式出现

The answer may be given in
another form

▲

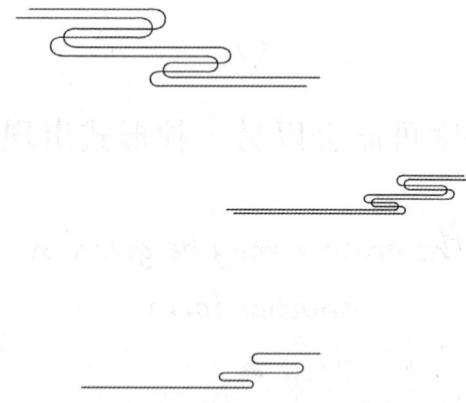

神奇的答案之书
THE BOOK OF MAGIC ANSWERS

▽

把它记下来

Write it down

▲

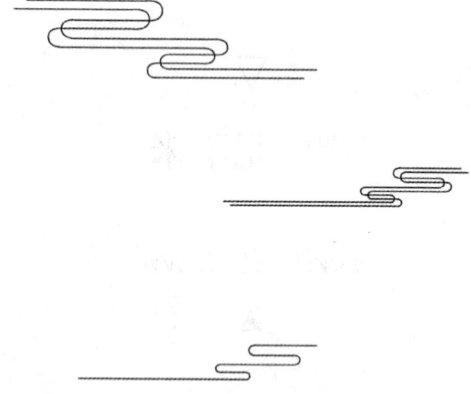

神奇的答案之书
THE BOOK OF MAGIC ANSWERS

了解得越透彻，你的目标便会越明确

If you understand more about it, you will know what to do more clearly

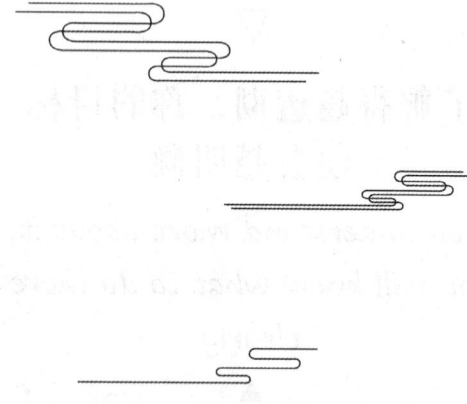

▽

需要做更多的努力

You need to make more effort

▲

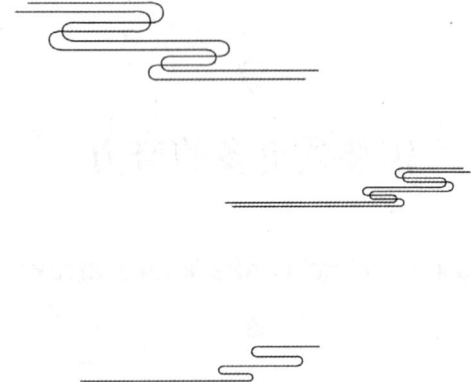

神奇的答案之书
THE BOOK OF MAGIC ANSWERS

▽

等待一个更好的机会

Wait for a better chance

▲

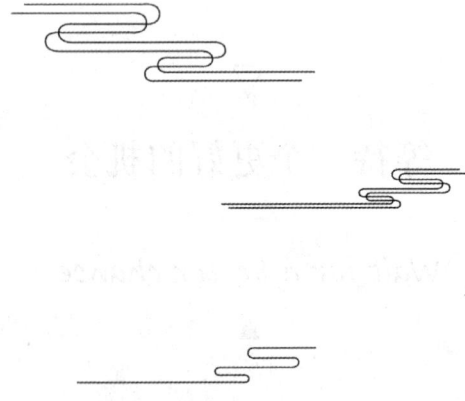

神奇的答案之书
THE BOOK OF MAGIC ANSWERS

神奇的答案之书
THE BOOK OF MAGIC ANSWERS

▽

数到十，再试一次

Count to ten and try again

▲

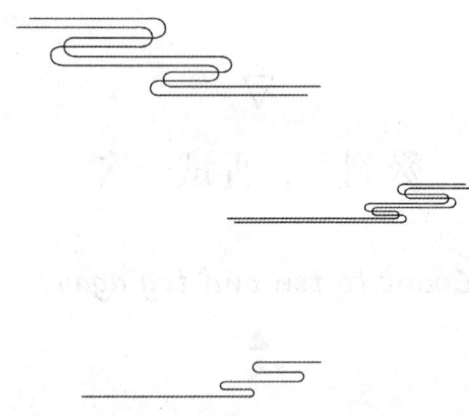

神奇的答案之书
THE BOOK OF MAGIC ANSWERS

▽

也许会很难，但值得

It may be very difficult, but it's worth it

▲

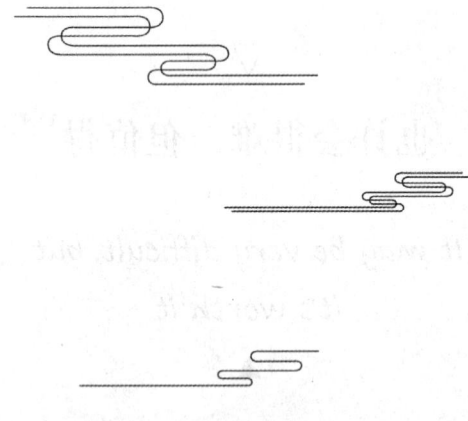

▽

你不得不放弃

You have to give it up

▲

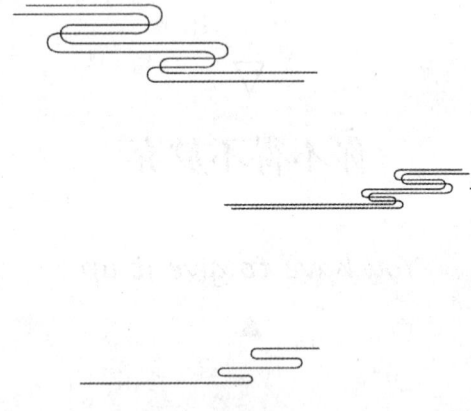

神奇的答案之书
THE BOOK OF MAGIC ANSWERS

神奇的答案之书
THE BOOK OF MAGIC ANSWERS

▽

很快就能决定

It will be decided soon

▲

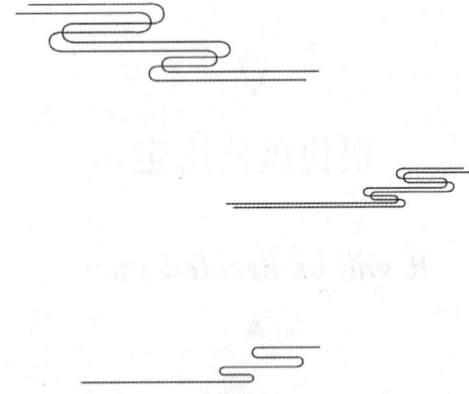

神奇的答案之书
THE BOOK OF MAGIC ANSWERS

▽

十分肯定

Pretty sure

▲

神奇的答案之书
THE BOOK OF MAGIC ANSWERS

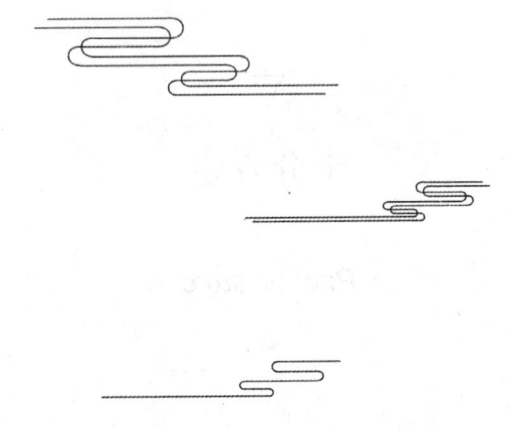

▽

你必须放弃

You must give it up

▲

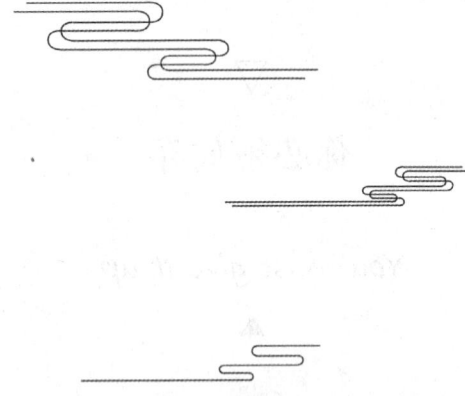

神奇的答案之书
THE BOOK OF MAGIC ANSWERS

▽

遵从规则

Follow the rules

▲

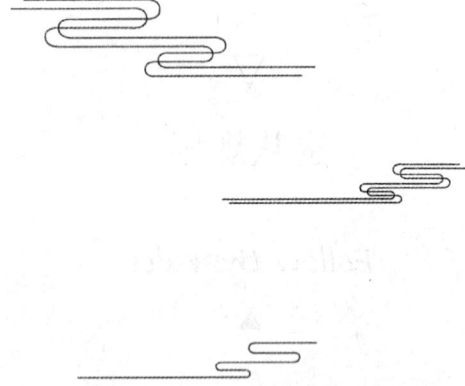

神奇的答案之书
THE BOOK OF MAGIC ANSWERS

▽

相关问题可能会出现

The related problems may appear

▲

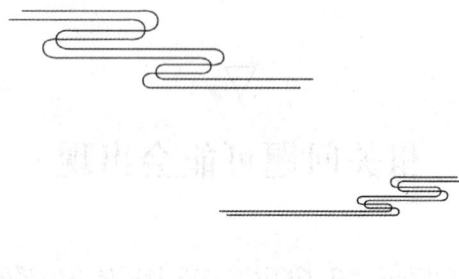

神奇的答案之书
THE BOOK OF MAGIC ANSWERS

▽

事情将遂你心意

Things will develop as you have expected

▲

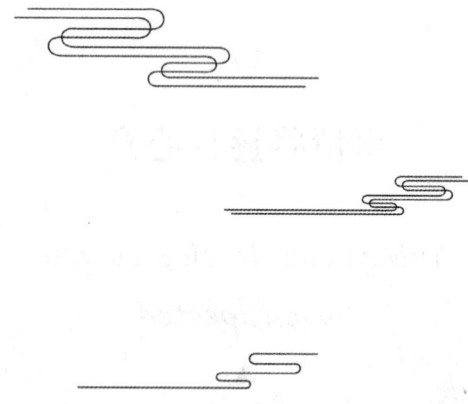

神奇的答案之书
THE BOOK OF MAGIC ANSWERS

▽

赌一把

Make a bet

▲

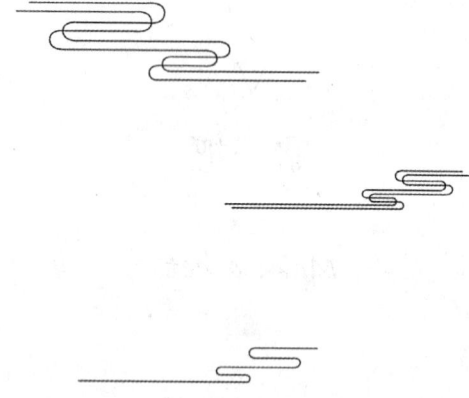

神奇的答案之书
THE BOOK OF MAGIC ANSWERS

神奇的答案之书
THE BOOK OF MAGIC ANSWERS

▽

以后再处理

Handle it later

▲

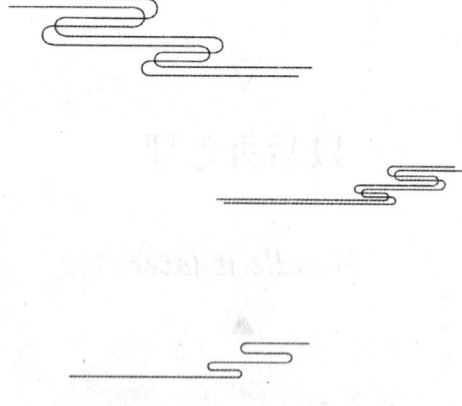

神奇的答案之书
THE BOOK OF MAGIC ANSWERS

结果将是乐观的

The result will be optimistic

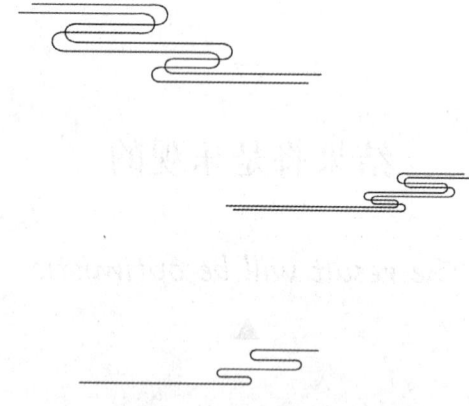

神奇的答案之书
THE BOOK OF MAGIC ANSWERS

▽

期待解决

Expect it to be resolved

▲

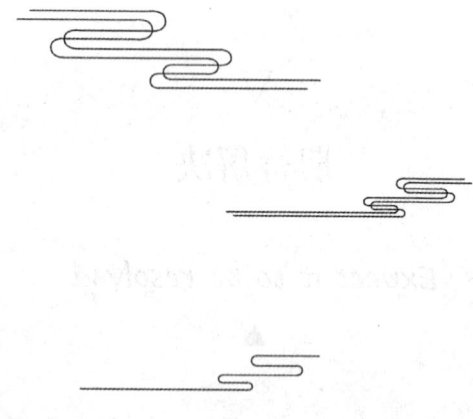

神奇的答案之书
THE BOOK OF MAGIC ANSWERS

把握契机，灵活应对

Grasp the chance and cope flexibly

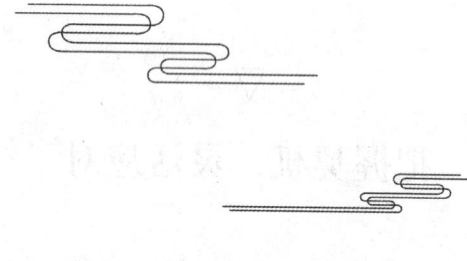

注意细节

Pay attention to details

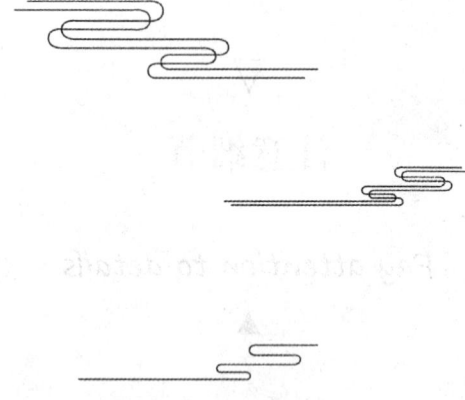

你的行动会使一切变好

Your action will make
everything get better

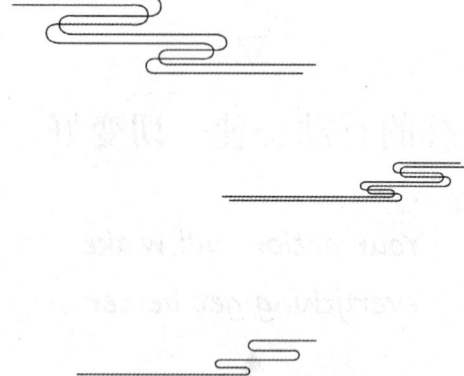

神奇的答案之书
THE BOOK OF MAGIC ANSWERS

神奇的答案之书
THE BOOK OF MAGIC ANSWERS

▽

清除你自身的短板

Eliminate your weakness

▲

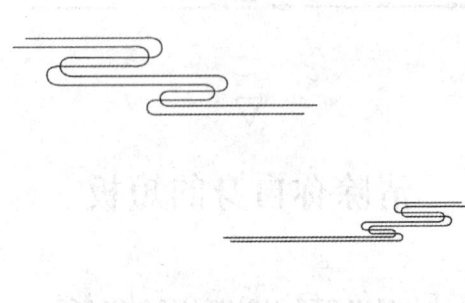
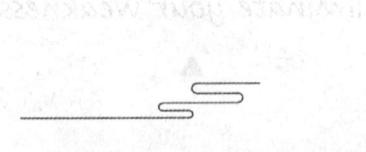

神奇的答案之书
THE BOOK OF MAGIC ANSWERS

▽

这是不可取的

This is not advisable

▲

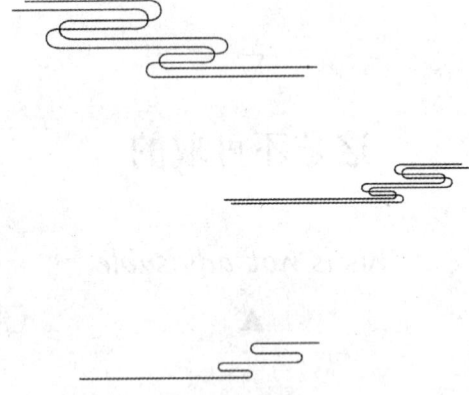

神奇的答案之书
THE BOOK OF MAGIC ANSWERS

神奇的答案之书
THE BOOK OF MAGIC ANSWERS

▽

将需要大量的努力

It needs a lot of effort

▲

是时候做打算了

It's time to make a plan

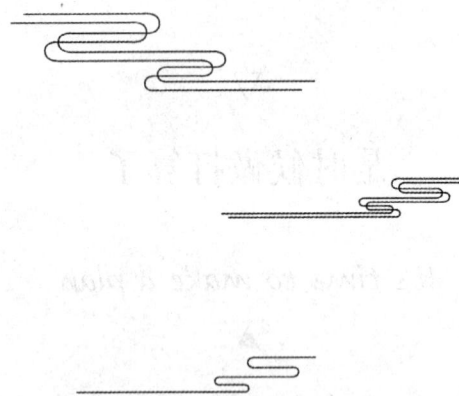

神奇的答案之书
THE BOOK OF MAGIC ANSWERS

▽

问问你最好的朋友

Ask your best friend

▲

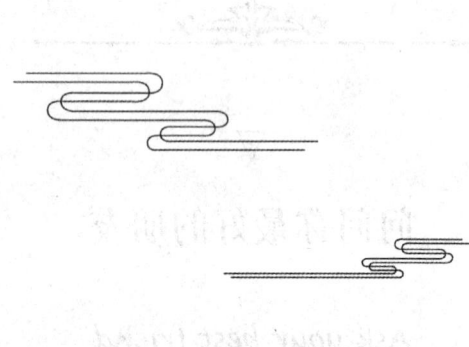

神奇的答案之书
THE BOOK OF MAGIC ANSWERS

享受这次体验

Enjoy this experience

神奇的答案之书
THE BOOK OF MAGIC ANSWERS

▽

要付出坚持不懈的努力

You need to make dogged effort

▲

你们还该不懈的努力

You need to make dogged efforts.

神奇的答案之书
THE BOOK OF MAGIC ANSWERS

▽

那仍旧无法预测

That can't be predicted yet

▲

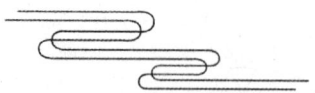
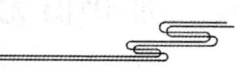
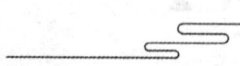

神奇的答案之书
THE BOOK OF MAGIC ANSWERS

▽

结果是乐观的

The result is optimistic

▲

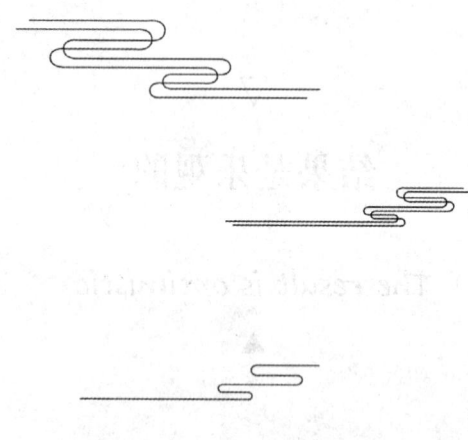

神奇的答案之书
THE BOOK OF MAGIC ANSWERS

▽

多花点时间来做决定

You should spend more time to
make a decision

▲

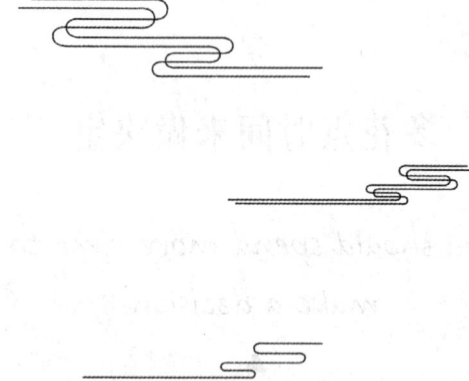

神奇的答案之书
THE BOOK OF MAGIC ANSWERS

▽

只做这一次

Do it just this once

▲

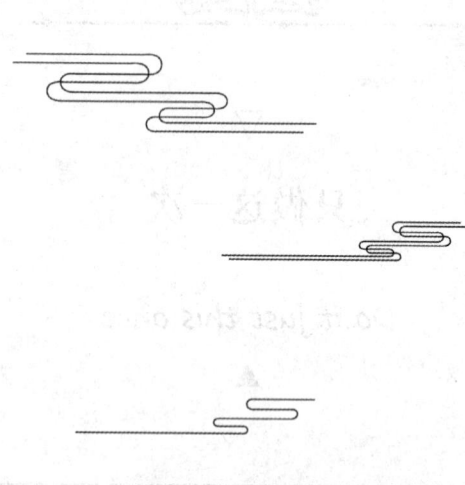

神奇的答案之书
THE BOOK OF MAGIC ANSWERS

▽

做些改变

Make changes

▲

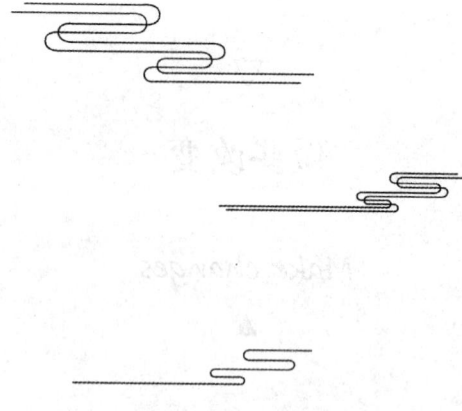

神奇的答案之书
THE BOOK OF MAGIC ANSWERS

▽

行得通

It is workable

▲

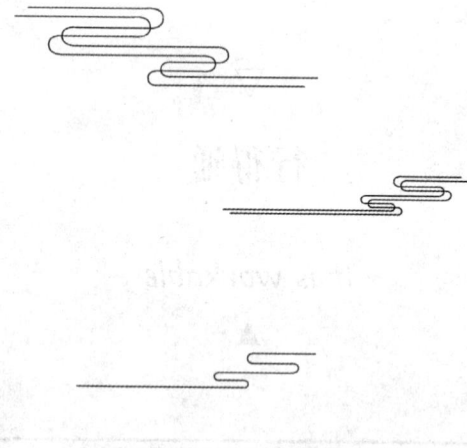

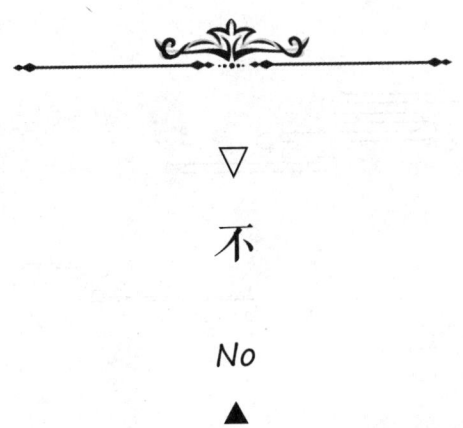

▽

不

No

▲

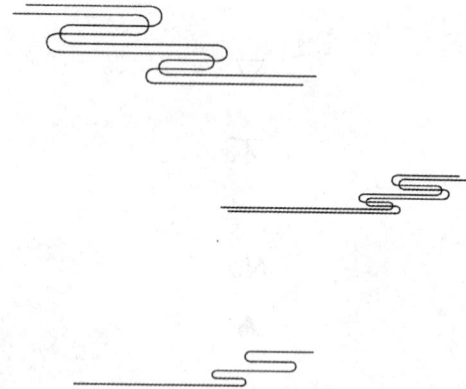

神奇的答案之书
THE BOOK OF MAGIC ANSWERS

神奇的答案之书
THE BOOK OF MAGIC ANSWERS

▽

相信你最开始的想法

Believe what you thought first

▲

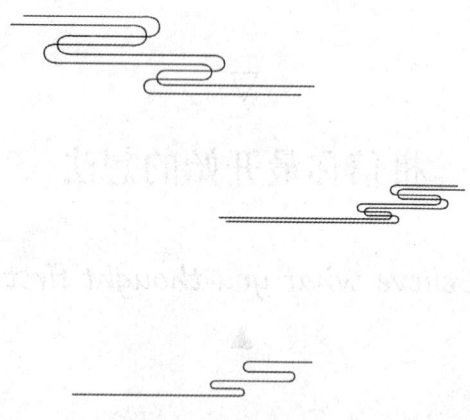

神奇的答案之书
THE BOOK OF MAGIC ANSWERS

▽

你需要适应

You need to adapt

▲

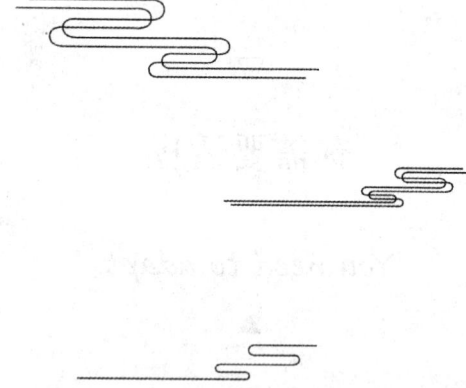

神奇的答案之书
THE BOOK OF MAGIC ANSWERS

▽

先主后次

Set up a priority catalogue in this process

▲

神奇的答案之书
THE BOOK OF MAGIC ANSWERS

$\triangledown$

这会让你付出代价

This will make you cost more

▲

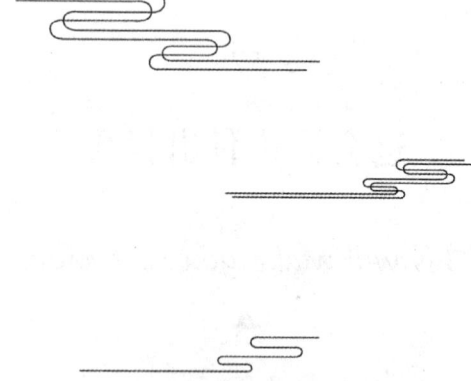

神奇的答案之书
THE BOOK OF MAGIC ANSWERS

神奇的答案之书
THE BOOK OF MAGIC ANSWERS

▽

尽早行动

Take action as early as possible

▲

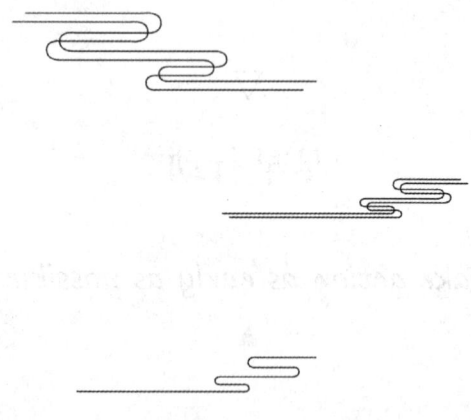

▽

可能会伤害到别人

It may do harm to others

▲

神奇的答案之书
THE BOOK OF MAGIC ANSWERS

▽

你比以往任何时候都更清楚

You know more clearly than ever before

▲

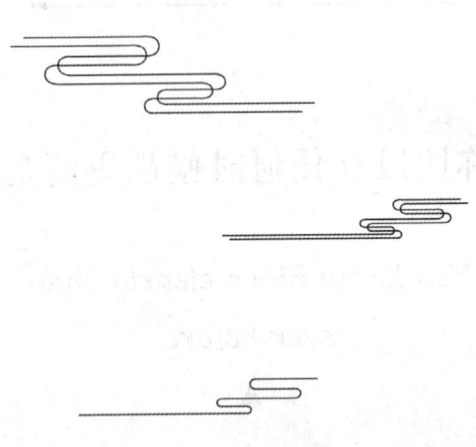

神奇的答案之书
THE BOOK OF MAGIC ANSWERS

极有可能发生变故

Something unexpected is most likely to happen

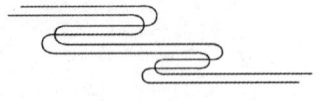
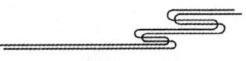
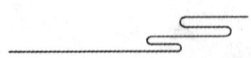

▽

毫无疑问

No doubt

▲

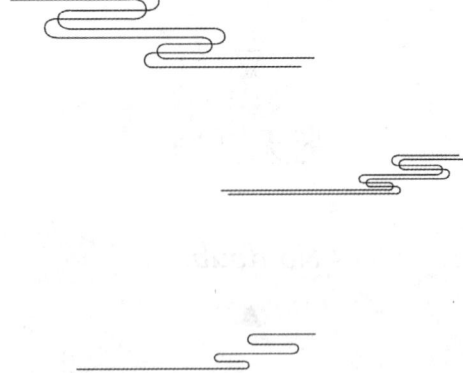

神奇的答案之书
THE BOOK OF MAGIC ANSWERS

▽

事情会朝目标发展

Things will go your way

▲

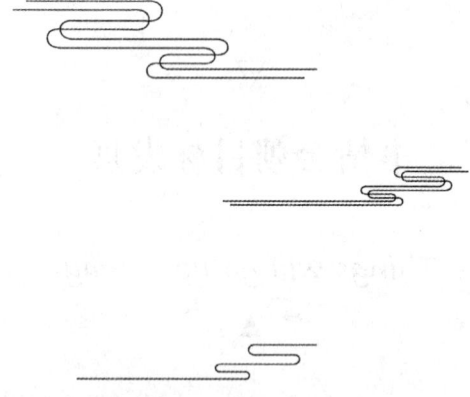

神奇的答案之书
THE BOOK OF MAGIC ANSWERS

▽

本页后第一个答案

The first answer after this page

▲

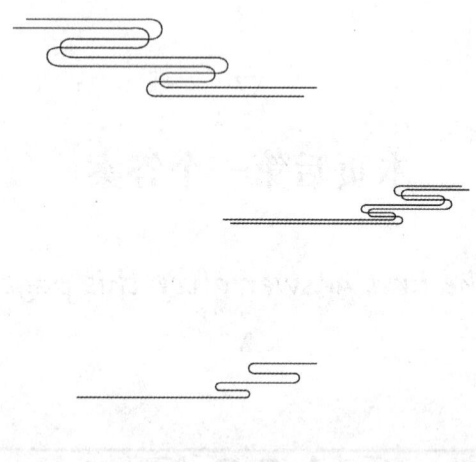

神奇的答案之书
THE BOOK OF MAGIC ANSWERS

列出这样做的理由

List the reason why it is done in this way

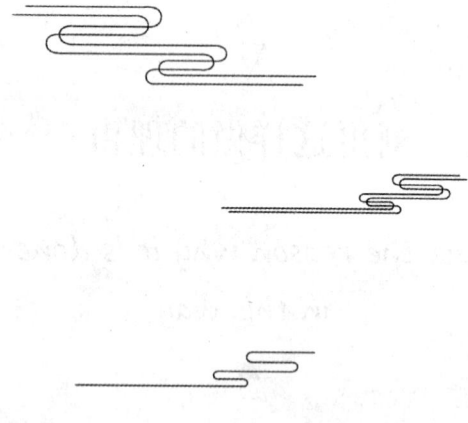

神奇的答案之书
THE BOOK OF MAGIC ANSWERS

▽

马上停下来

Stop at once

▲

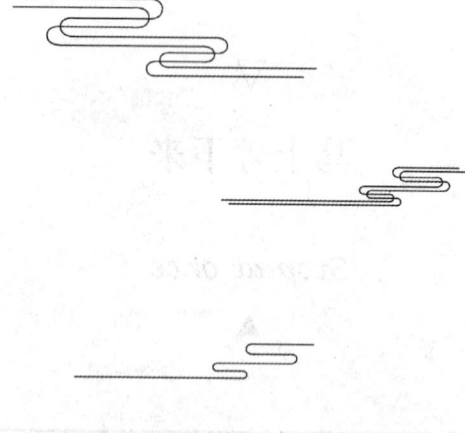

▽

不是很确定

It is not sure

▲

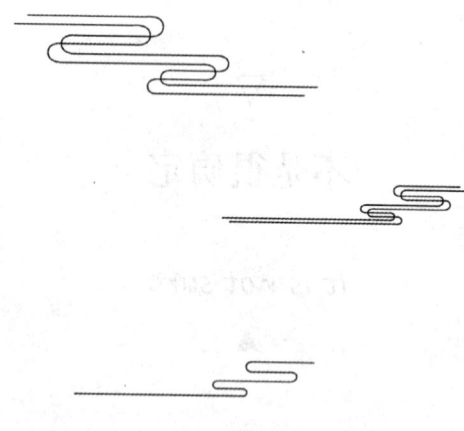

神奇的答案之书
THE BOOK OF MAGIC ANSWERS

▽

不用担心

Don't worry

▲

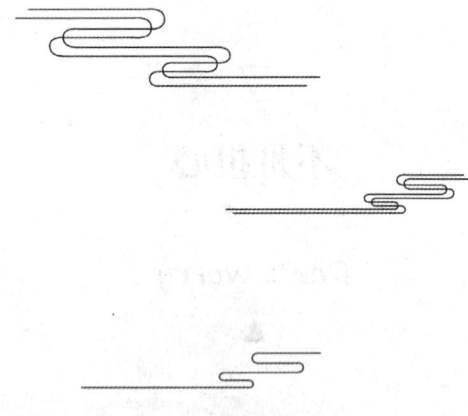

▽

首先做好自己的事

First get your own things done

▲

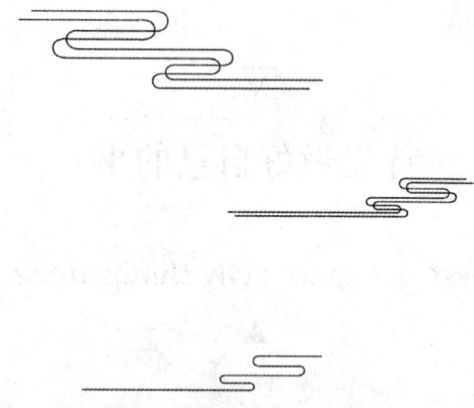

神奇的答案之书
THE BOOK OF MAGIC ANSWERS

神奇的答案之书
THE BOOK OF MAGIC ANSWERS

▽

情况很快就会有变化

The situation will change soon

▲

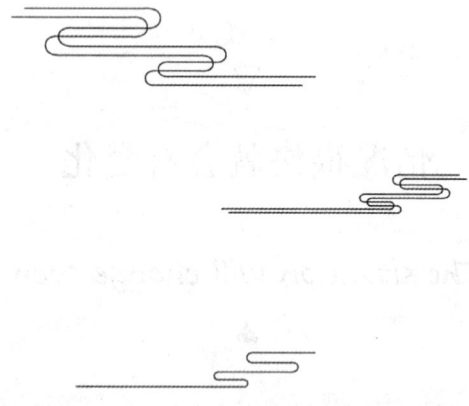

神奇的答案之书
THE BOOK OF MAGIC ANSWERS

▽

不要告诉别人

Don't tell others

▲

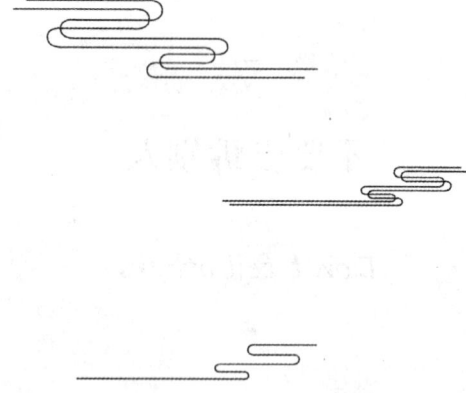

神奇的答案之书
THE BOOK OF MAGIC ANSWERS

▽

你需要其他人的帮助

You need help from others

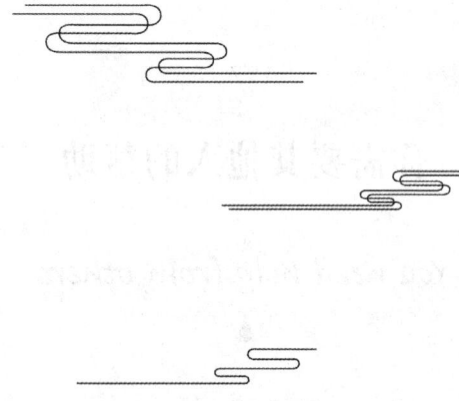

神奇的答案之书
THE BOOK OF MAGIC ANSWERS

▽

那是一件乐事

That is a pleasure

▲

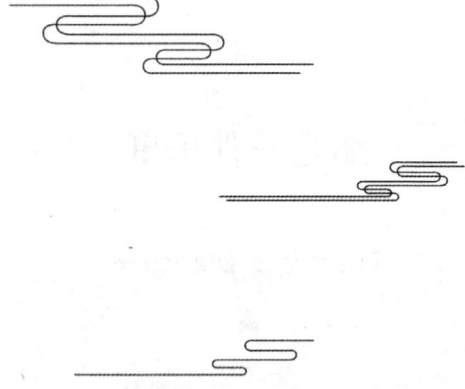

神奇的答案之书
THE BOOK OF MAGIC ANSWERS

▽

现在你就行

You can manage it now

▲

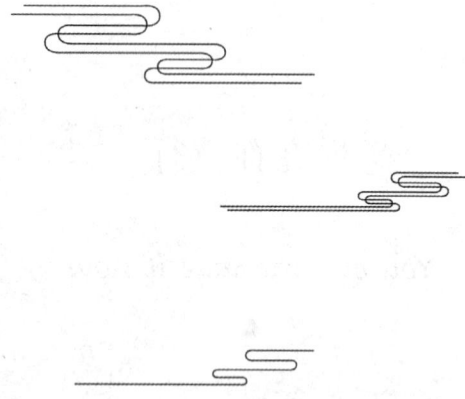

神奇的答案之书
THE BOOK OF MAGIC ANSWERS

▽

仍然无法预测

This could not be predicted yet

▲

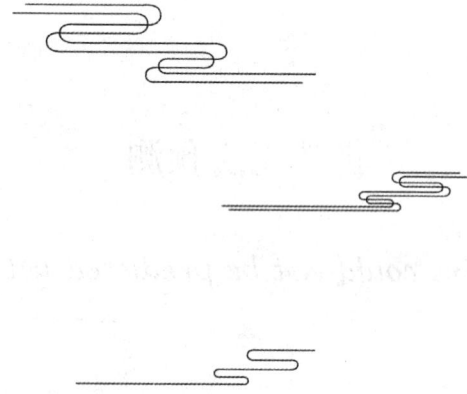

神奇的答案之书
THE BOOK OF MAGIC ANSWERS

▽

列出理由

List reasons

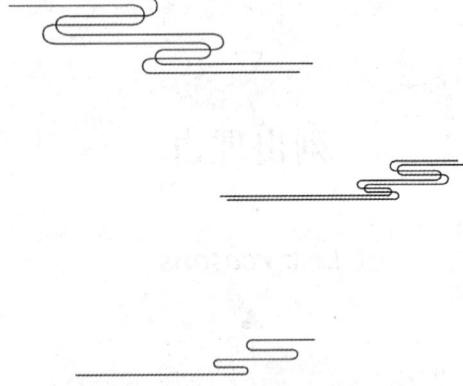

神奇的答案之书
THE BOOK OF MAGIC ANSWERS

压力之下不要草率行事

Don't do things recklessly when under pressure

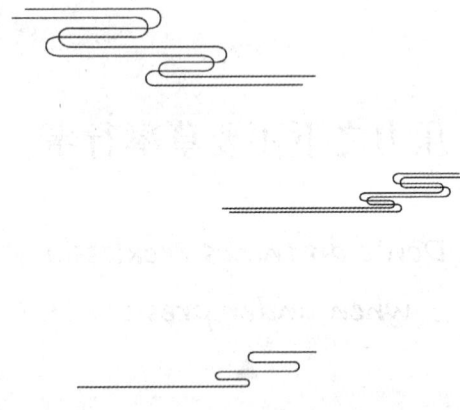

神奇的答案之书
THE BOOK OF MAGIC ANSWERS

▽

你需要主动一些

You need to behave actively

▲

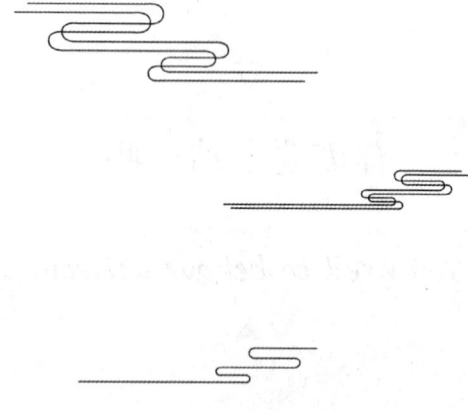

神奇的答案之书
THE BOOK OF MAGIC ANSWERS

▽

不要等待

Don't wait

▲

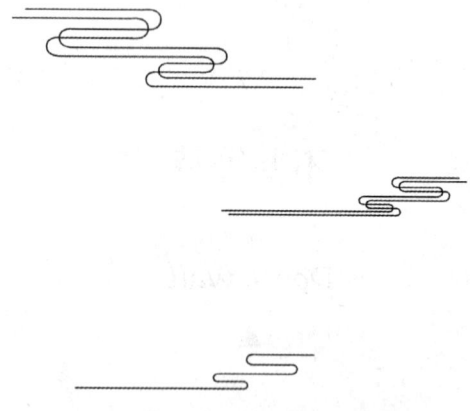

神奇的答案之书
THE BOOK OF MAGIC ANSWERS

▽

是

Yes

▲

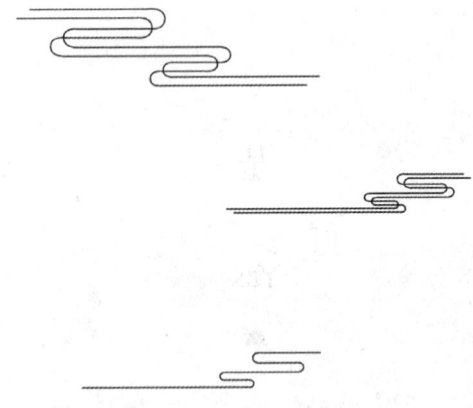

神奇的答案之书
THE BOOK OF MAGIC ANSWERS

▽

你必须马上行动

You must take action at once

▲

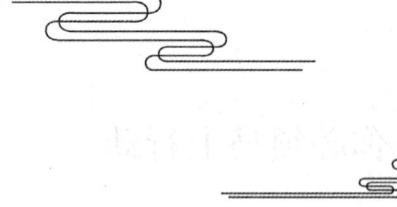

神奇的答案之书
THE BOOK OF MAGIC ANSWERS

▽

这并不重要

It's not important

▲

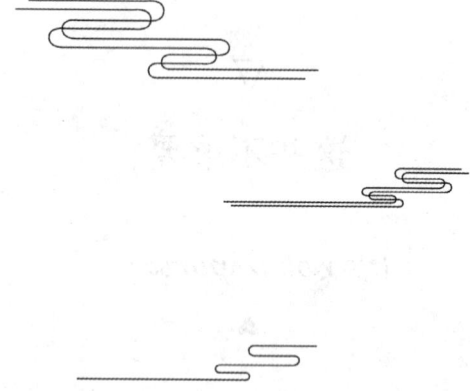

神奇的答案之书
THE BOOK OF MAGIC ANSWERS

告诉别人这对你的意义

Tell someone what it means to you

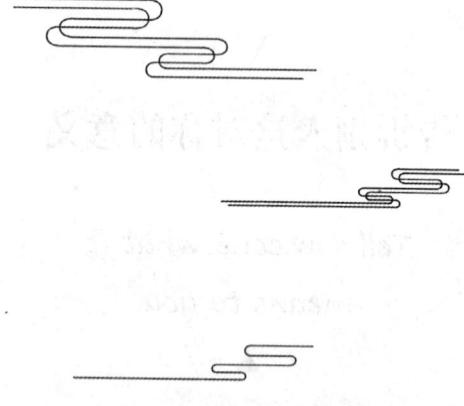

神奇的答案之书
THE BOOK OF MAGIC ANSWERS

▽

保持开放的心态

Keep an open mind

▲

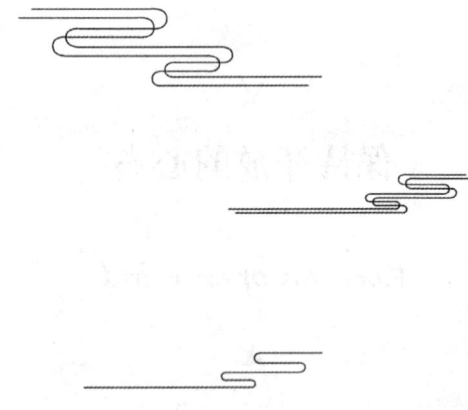

神奇的答案之书
THE BOOK OF MAGIC ANSWERS

▽

这是制订计划的好时机

It's good time to make plans

▲

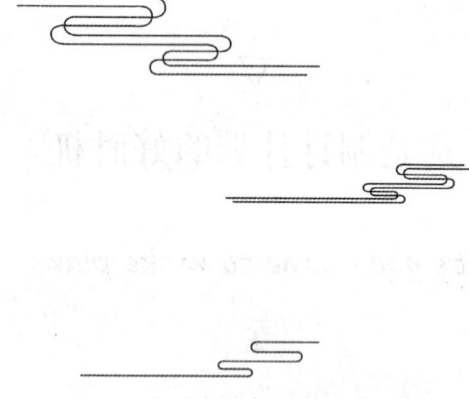

神奇的答案之书
THE BOOK OF MAGIC ANSWERS

这可能很困难，
但你会发现它的价值

It may be difficult, but you will find value in it

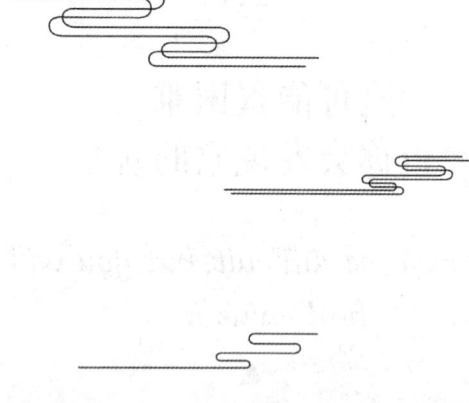

神奇的答案之书
THE BOOK OF MAGIC ANSWERS

这点麻烦是值得的

It's worth the trouble

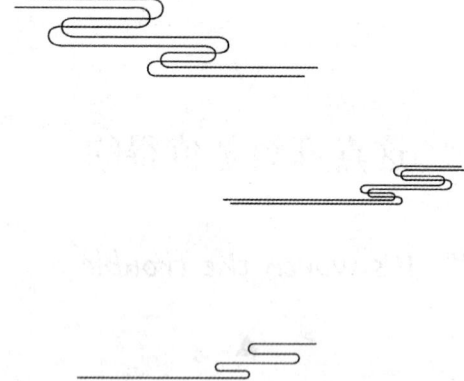

神奇的答案之书
THE BOOK OF MAGIC ANSWERS

◇

▽

你肯定会被支持的

You will be supported certainly

▲

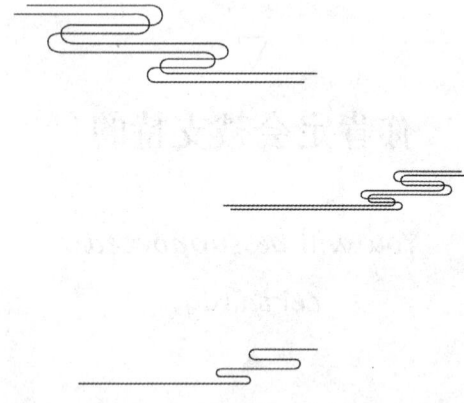

神奇的答案之书
THE BOOK OF MAGIC ANSWERS

▽

合作是关键

The key is cooperation

▲

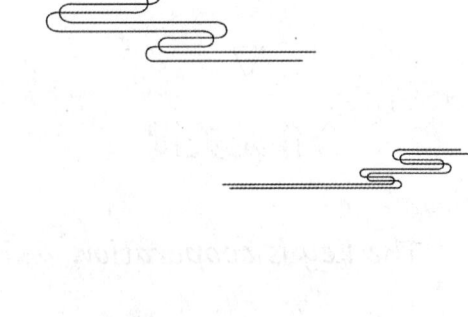

神奇的答案之书
THE BOOK OF MAGIC ANSWERS

▽

寻求更多的选择

Seek more options

▲

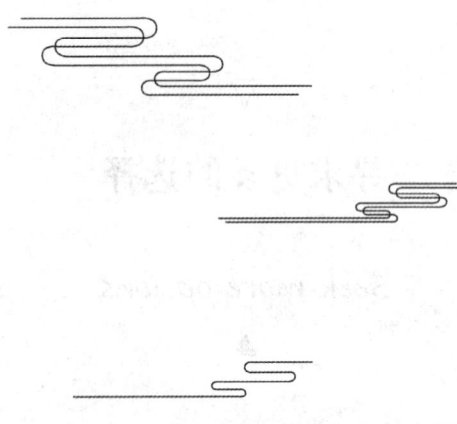

神奇的答案之书
THE BOOK OF MAGIC ANSWERS

▽

你不可能失败

You can't fail

▲

神奇的答案之书
THE BOOK OF MAGIC ANSWERS

▽

你必须现在行动

You must act now

▲

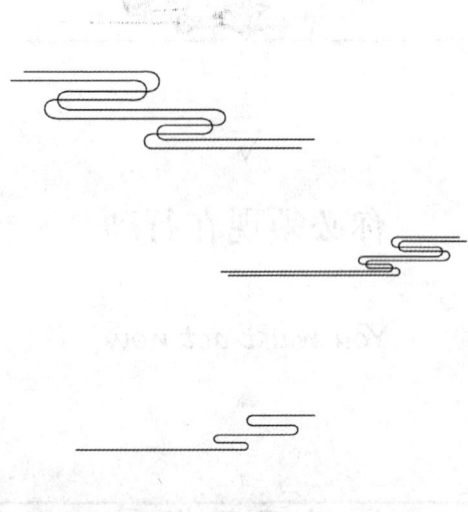

神奇的答案之书
THE BOOK OF MAGIC ANSWERS

坚持就是胜利

Success belongs to the persevering

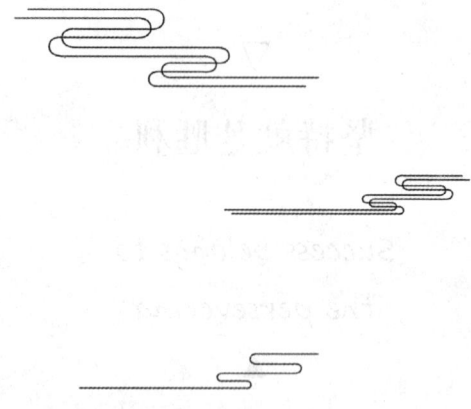

神奇的答案之书
THE BOOK OF MAGIC ANSWERS

▽

你不会失望

You will not be disappointed

▲

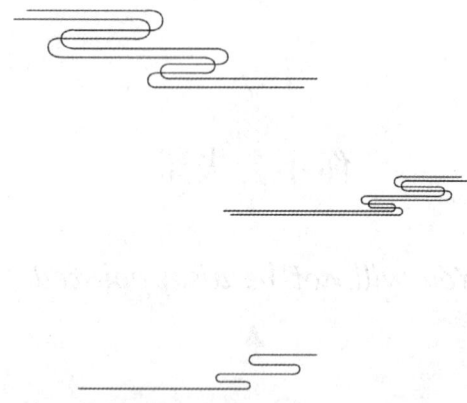

神奇的答案之书
THE BOOK OF MAGIC ANSWERS

▽

遵从你的意愿

Follow your good intentions

▲

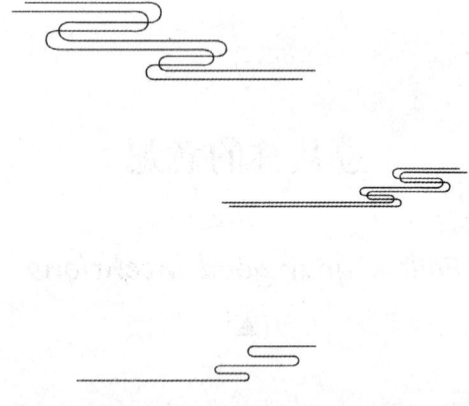

神奇的答案之书
THE BOOK OF MAGIC ANSWERS

花更多的时间来决定

Spend more time to decide

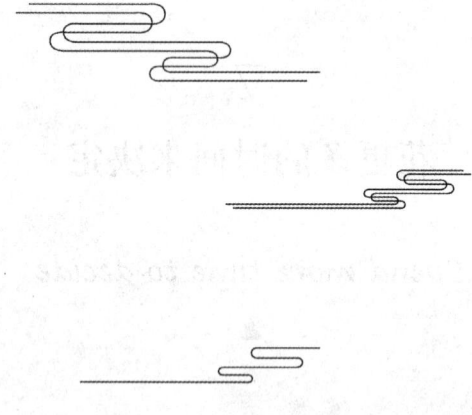

神奇的答案之书
THE BOOK OF MAGIC ANSWERS

别忽视显而易见的东西

Don't ignore the obvious things

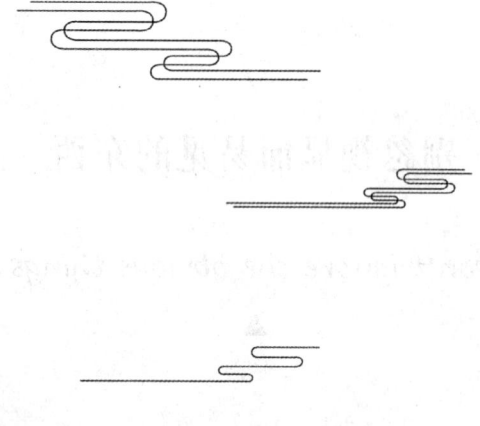

神奇的答案之书
THE BOOK OF MAGIC ANSWERS

▽

相信你最初的想法

Believe what you first thought

▲

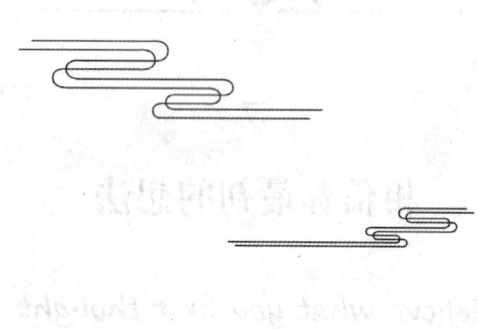

相信你看到的想法

Believe what you first thought.

神奇的答案之书
THE BOOK OF MAGIC ANSWERS

▽

请不要抗拒

Please don't resist

▲

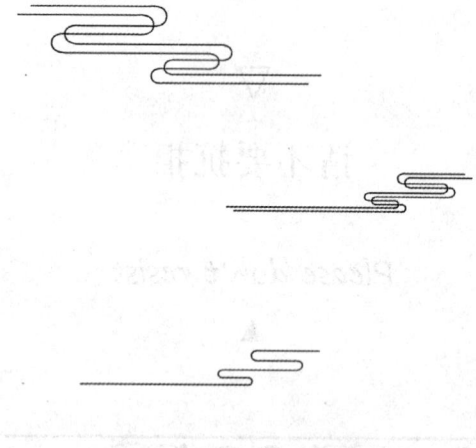

▽

这带来好运

That will bring you good luck

▲

神奇的答案之书
THE BOOK OF MAGIC ANSWERS

▽

不要迫于压力草率行事

*Don't behave recklessly
when under pressure*

▲

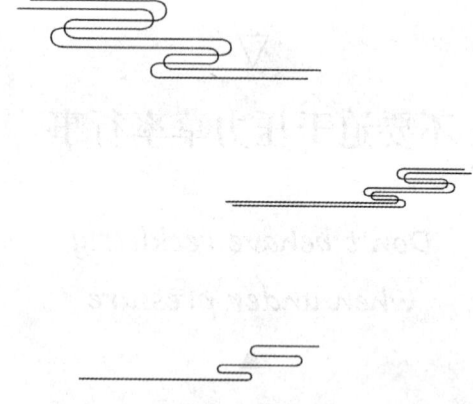

可能发生小意外

Some little things may happen unexpectedly

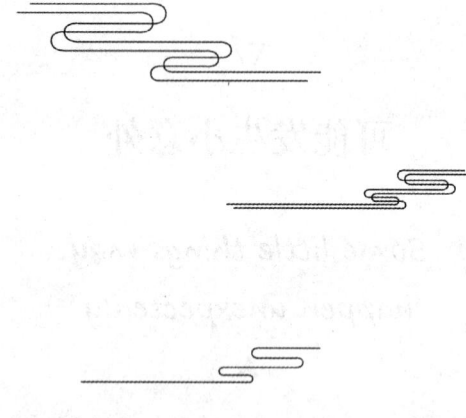

神奇的答案之书
THE BOOK OF MAGIC ANSWERS

神奇的答案之书
THE BOOK OF MAGIC ANSWERS

研究并享受它

Study and enjoy it

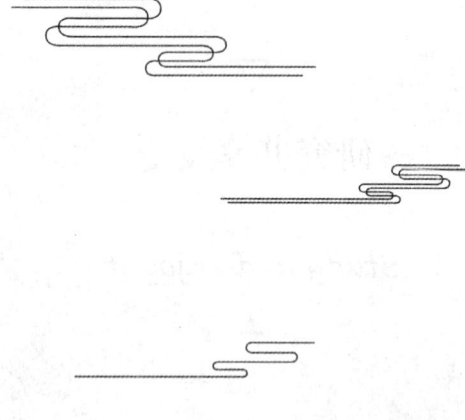

神奇的答案之书
THE BOOK OF MAGIC ANSWERS

▽

不要妄下赌注

Don't make a bet wantonly

▲

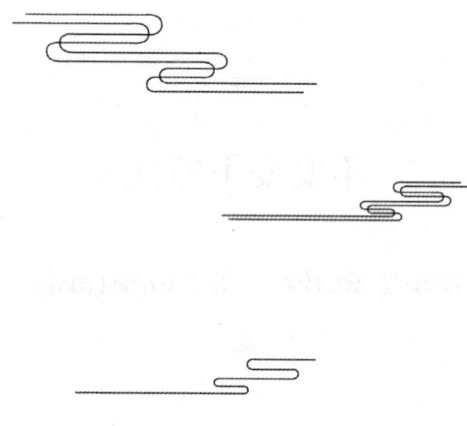

神奇的答案之书
THE BOOK OF MAGIC ANSWERS

神奇的答案之书
THE BOOK OF MAGIC ANSWERS

▽

接受改变

Accept changes

▲

○●○●●●○●○○○●●●

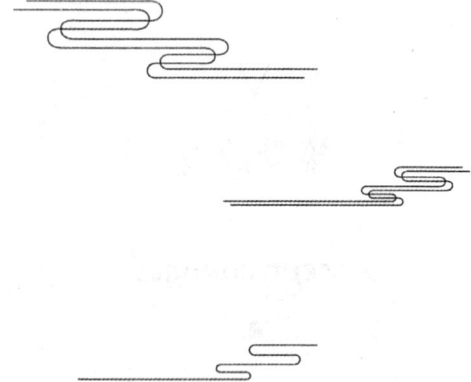

神奇的答案之书
THE BOOK OF MAGIC ANSWERS

可能已经无法改变

It perhaps can not be changed anymore

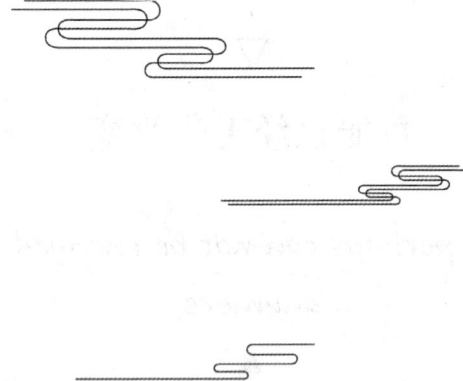

神奇的答案之书
THE BOOK OF MAGIC ANSWERS

一些援助能确保你成功

Some assistance can ensure your success

神奇的答案之书
THE BOOK OF MAGIC ANSWERS

▽

这也取决于另一种情况

That also depends on another situation

▲

神奇的答案之书
THE BOOK OF MAGIC ANSWERS

神奇的答案之书
THE BOOK OF MAGIC ANSWERS

▽

你肯定会获得支持

You will surely get the support

▲

▽

仅做一次

Do it only once

▲

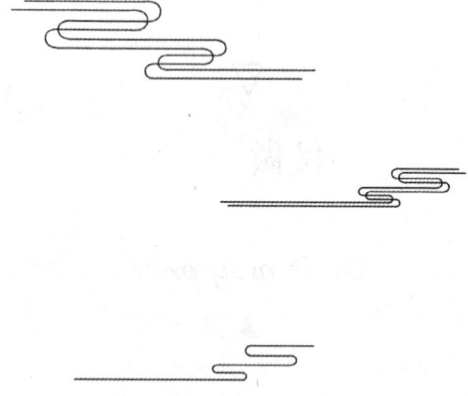

神奇的答案之书
THE BOOK OF MAGIC ANSWERS

听从朋友的建议

Follow the advice of your friend

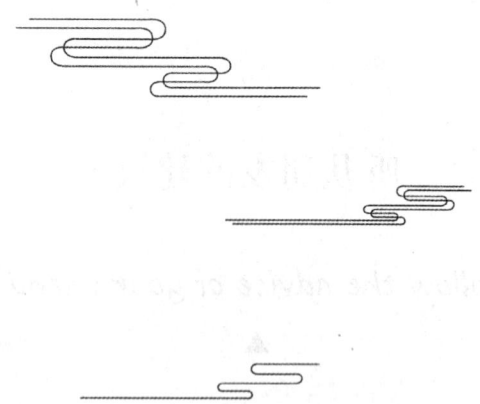

神奇的答案之书
THE BOOK OF MAGIC ANSWERS

▽

如你所愿

Things will get on just as you wish

▲

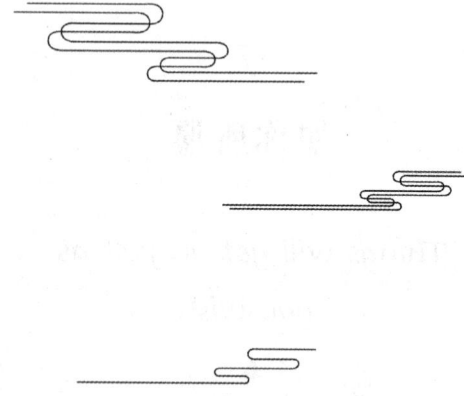

神奇的答案之书
THE BOOK OF MAGIC ANSWERS

神奇的答案之书
THE BOOK OF MAGIC ANSWERS

▽

当局者迷

Those closely involved cannot see clearly

▲

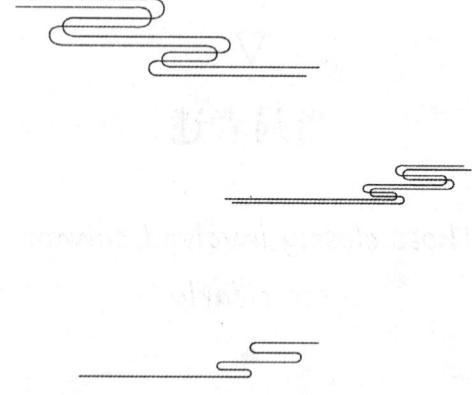

神奇的答案之书
THE BOOK OF MAGIC ANSWERS

▽

无论你怎么做，结果依旧

Whatever you do, the result is the same

▲

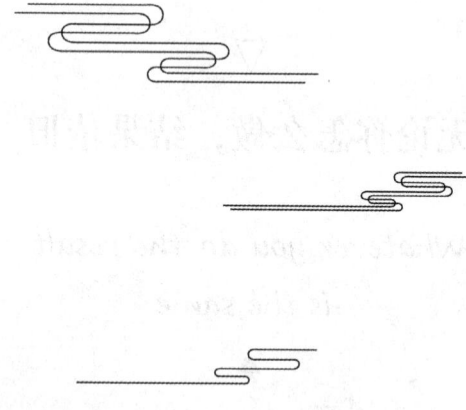

神奇的答案之书
THE BOOK OF MAGIC ANSWERS

▽

问问你的异性同事

Ask your colleagues of
the opposite sex

▲

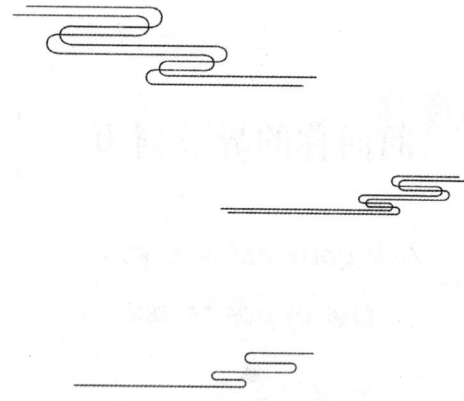

神奇的答案之书
THE BOOK OF MAGIC ANSWERS

▽

履行你的义务

Perform your duty

▲

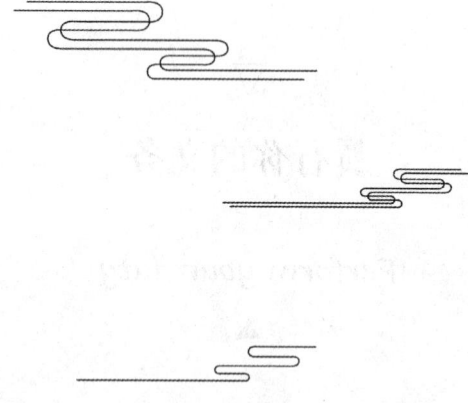

神奇的答案之书
THE BOOK OF MAGIC ANSWERS

你能以任何方式改善现状

You can improve the present situation in any way

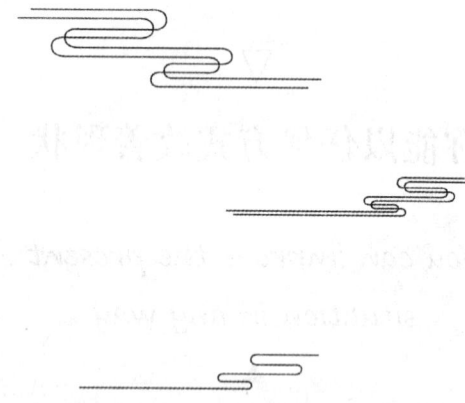

神奇的答案之书
THE BOOK OF MAGIC ANSWERS

▽

为了做出最好的选择，
请保持冷静

In order to make the best choice,
please keep calm

▲

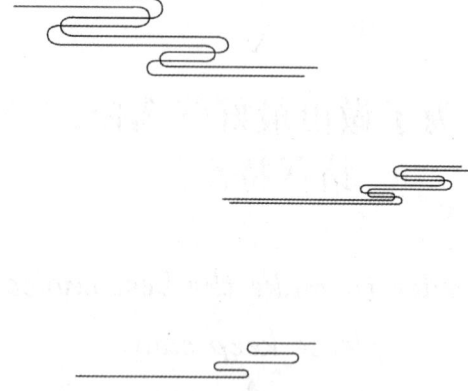

神奇的答案之书
THE BOOK OF MAGIC ANSWERS

你会为此感到高兴

You will be glad for it

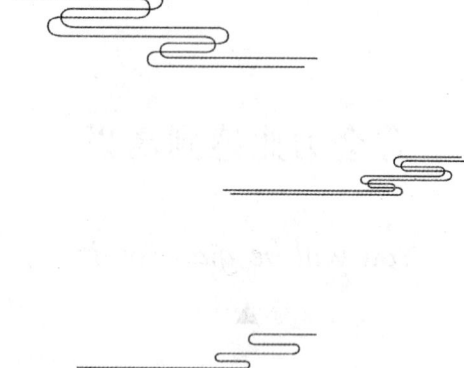

神奇的答案之书
THE BOOK OF MAGIC ANSWERS

放弃之前的想法

Give up the ideas you held before

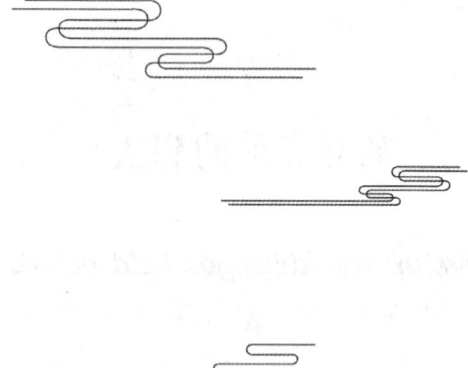

神奇的答案之书
THE BOOK OF MAGIC ANSWERS

▽

顺从你的意愿

Abide by your wishes

▲

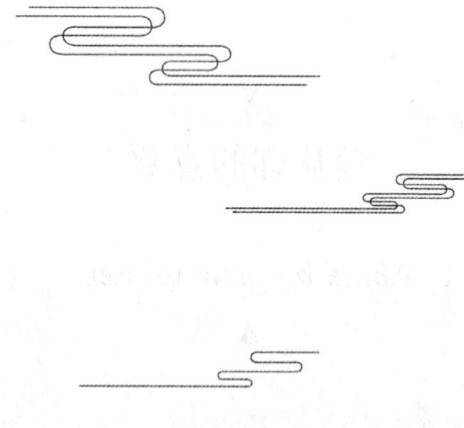

神奇的答案之书
THE BOOK OF MAGIC ANSWERS

▽

你不会忘记这些

You will not forget about these

▲

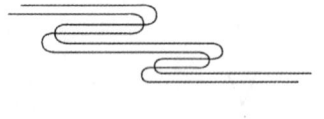
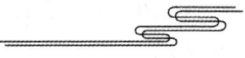

神奇的答案之书
THE BOOK OF MAGIC ANSWERS

▽

不要在乎

Don't care about it

▲

神奇的答案之书
THE BOOK OF MAGIC ANSWERS

一些援助有利于你成功

Some assistance can ensure your success

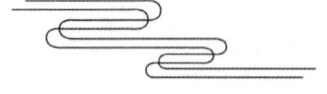
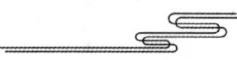
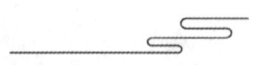

神奇的答案之书
THE BOOK OF MAGIC ANSWERS

▽

谨慎对待

Deal with it with discretion

▲

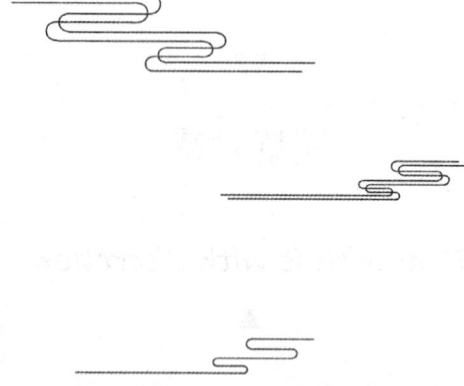

神奇的答案之书
THE BOOK OF MAGIC ANSWERS

神奇的答案之书
THE BOOK OF MAGIC ANSWERS

▽

放弃你现在的想法

Give up your current idea

▲

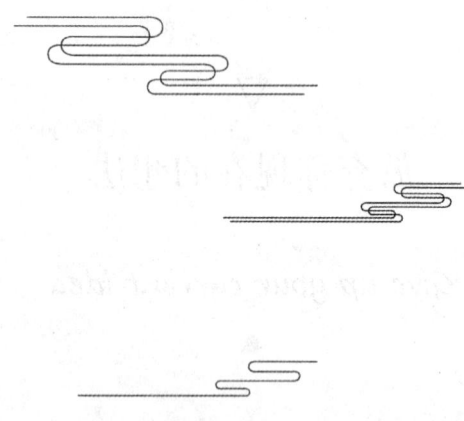

神奇的答案之书
THE BOOK OF MAGIC ANSWERS

▽

有理由保持乐观

There are reasons to be optimistic

▲

神奇的答案之书
THE BOOK OF MAGIC ANSWERS

▽

带着好奇去探索

To explore with curiosity

▲

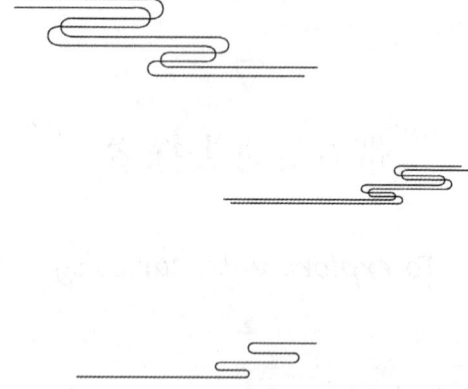

神奇的答案之书
THE BOOK OF MAGIC ANSWERS

神奇的答案之书
THE BOOK OF MAGIC ANSWERS

▽
你会发现自己难以妥协

You will find it difficult to
compromise

▲

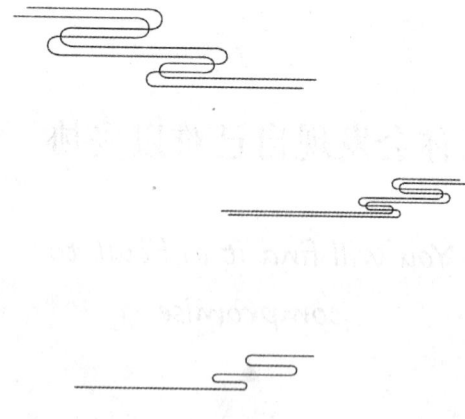

神奇的答案之书
THE BOOK OF MAGIC ANSWERS

▽

是时候做新打算了

It's time to make new plans

▲

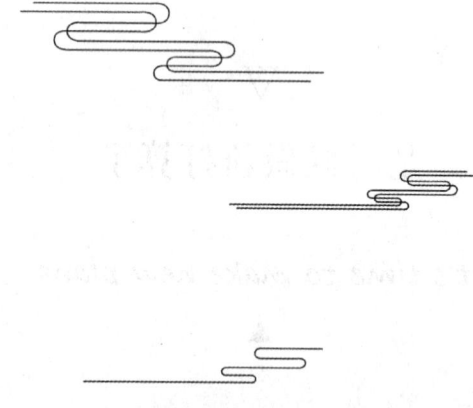

▽

改变不会很快发生

The change will not come soon

▲

神奇的答案之书
THE BOOK OF MAGIC ANSWERS

仔细地聆听，
然后你就会知道

神奇的答案之书
THE BOOK OF MAGIC ANSWERS

▽

耐心点

Be patient

▲

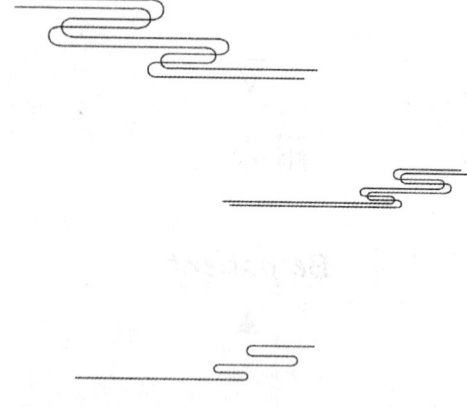

神奇的答案之书
THE BOOK OF MAGIC ANSWERS

▽

果断放弃

Give it up firmly

▲

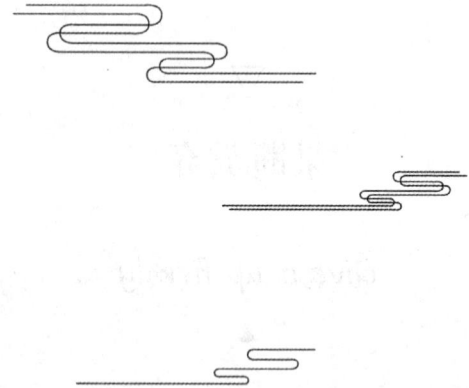

神奇的答案之书
THE BOOK OF MAGIC ANSWERS

最好把心思放在工作上

You'd better keep your mind on your work

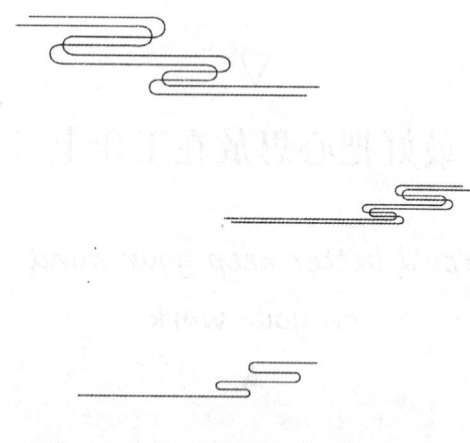

要做就做好，
否则就不要去做

Don't do it or do it best

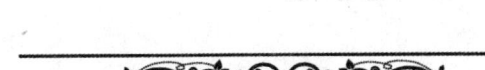

神奇的答案之书
THE BOOK OF MAGIC ANSWERS

▽

深表怀疑

Seriously doubt it

▲

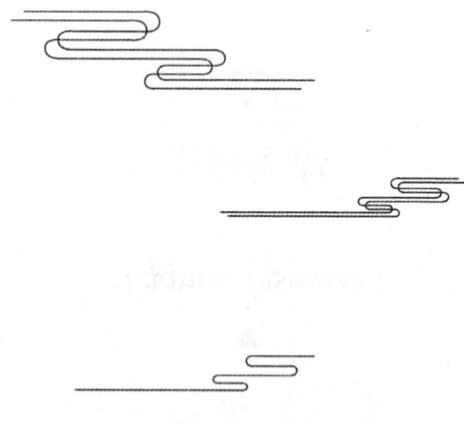

神奇的答案之书
THE BOOK OF MAGIC ANSWERS

▽

运用你的想象力

Use your imagination

▲

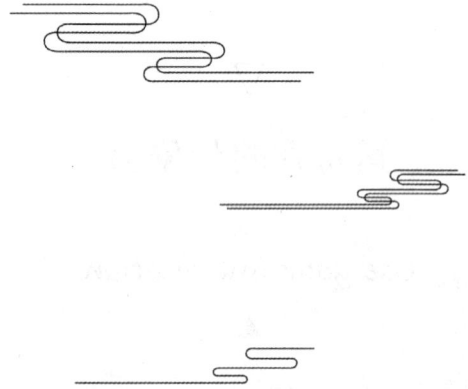

神奇的答案之书
THE BOOK OF MAGIC ANSWERS

已超出你的控制

Beyond your control

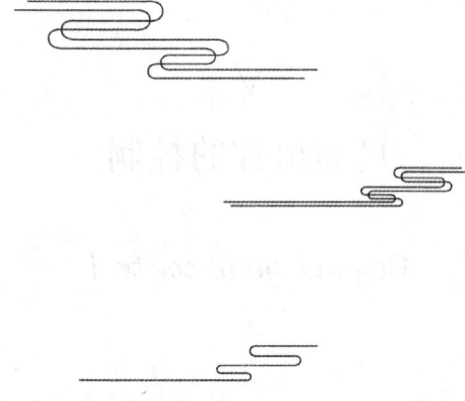

神奇的答案之书
THE BOOK OF MAGIC ANSWERS

▽

你已然发现自己无法妥协

You have already found yourself
unable to compromise

▲

神奇的答案之书
THE BOOK OF MAGIC ANSWERS

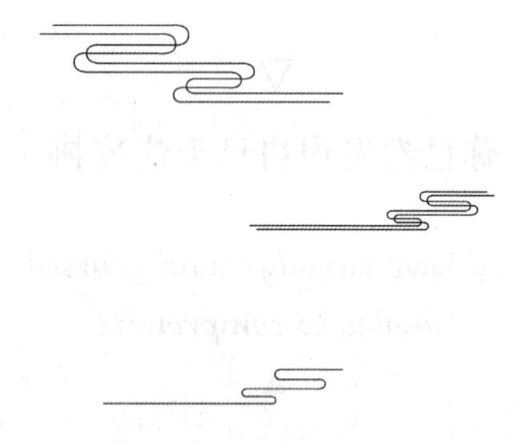

你需要更多信息

You need more information

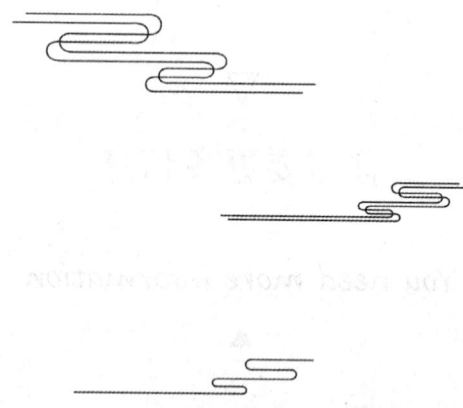

神奇的答案之书
THE BOOK OF MAGIC ANSWERS

数到五,再问一次

Count to five and ask again

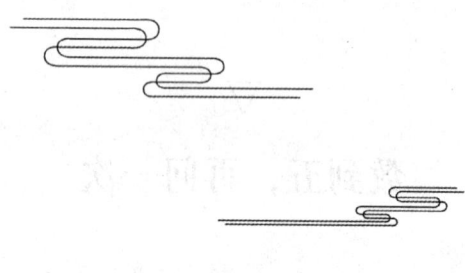
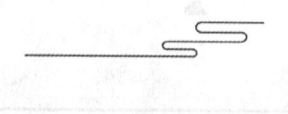

神奇的答案之书
THE BOOK OF MAGIC ANSWERS

▽

开阔视野

Broaden your vision

▲

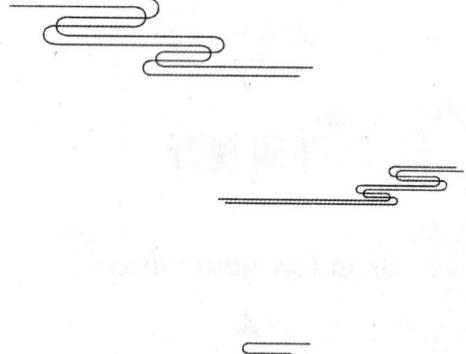

神奇的答案之书
THE BOOK OF MAGIC ANSWERS

▽

明天再试试

Try again tomorrow

▲

神奇的答案之书
THE BOOK OF MAGIC ANSWERS

注意细节

Pay attention to details

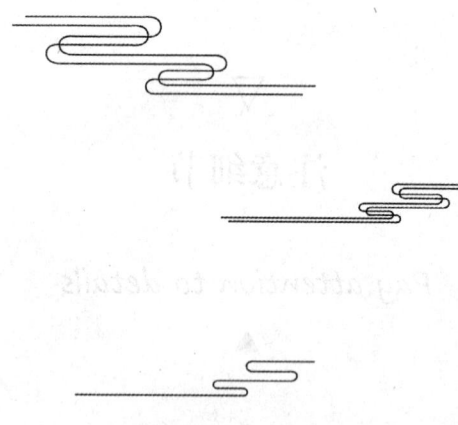

花絮扣扣
Fag attraction to details

神奇的答案之书
THE BOOK OF MAGIC ANSWERS

▽

看得更清楚些

See more clearly

▲

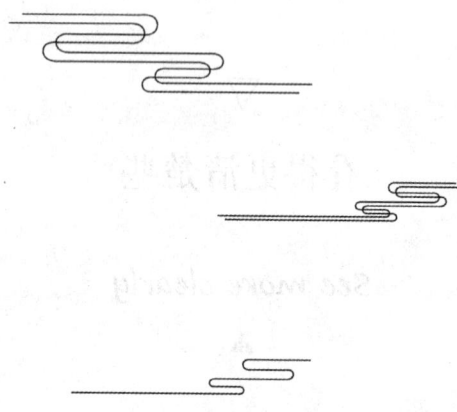

神奇的答案之书
THE BOOK OF MAGIC ANSWERS

神奇的答案之书
THE BOOK OF MAGIC ANSWERS

▽

结果可能会令人震惊

The result may be sensational

▲

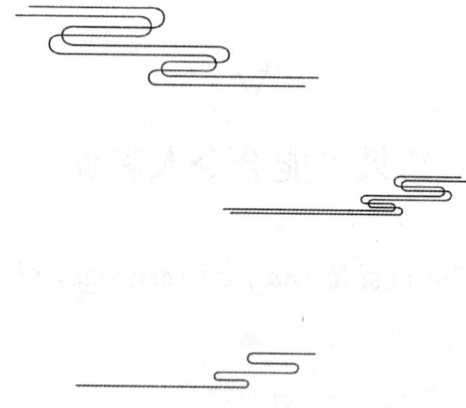

神奇的答案之书
THE BOOK OF MAGIC ANSWERS

▽

不要浪费你的精力

Don't dissipate your effort

▲

▽

管它呢

Whatever

▲

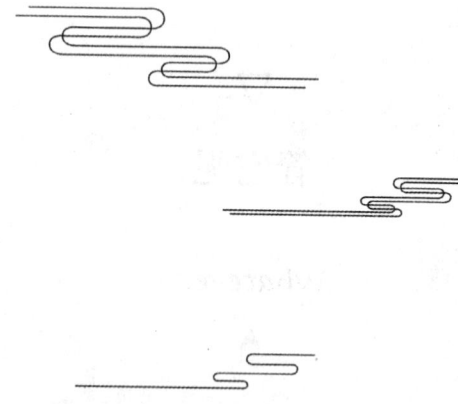

神奇的答案之书
THE BOOK OF MAGIC ANSWERS

神奇的答案之书
THE BOOK OF MAGIC ANSWERS

▽

发挥你的想象力

Free your imagination

▲

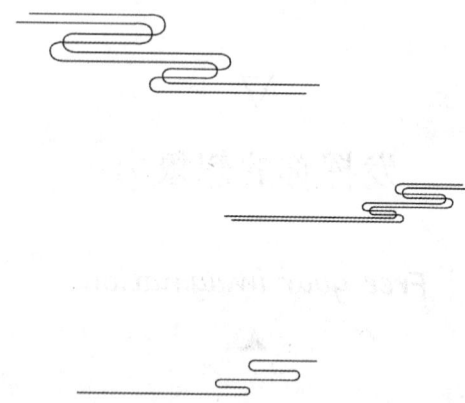

神奇的答案之书
THE BOOK OF MAGIC ANSWERS

那将引起一些纷争

That will cause some disputes

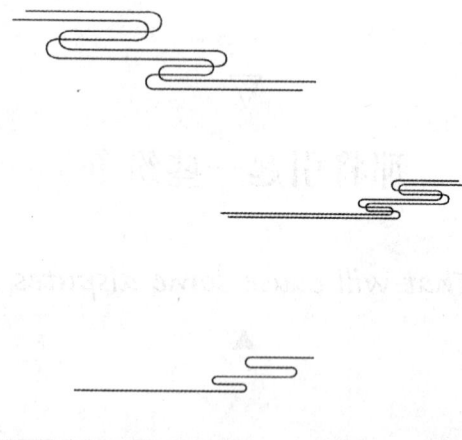

神奇的答案之书
THE BOOK OF MAGIC ANSWERS

▽

相关问题可能会浮出水面

The related problems will rise to the surface

▲

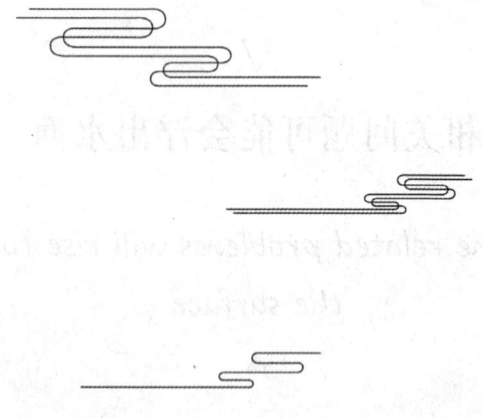

神奇的答案之书
THE BOOK OF MAGIC ANSWERS

先主后次

Set up a priority catalogue
in this process

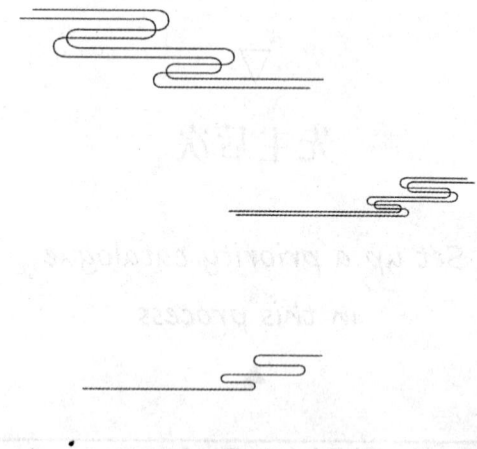

▽

重新考虑下你的方法

Make a reconsideration about
your method

▲

神奇的答案之书
THE BOOK OF MAGIC ANSWERS

神奇的答案之书
THE BOOK OF MAGIC ANSWERS

▽

你必须随机应变

You must be flexible

▲

▽

你可能遭遇反对

You may meet with opposition

▲

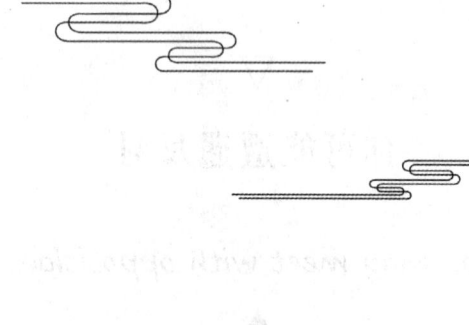

▽

你能否不要抗拒

Can you just not resist

▲

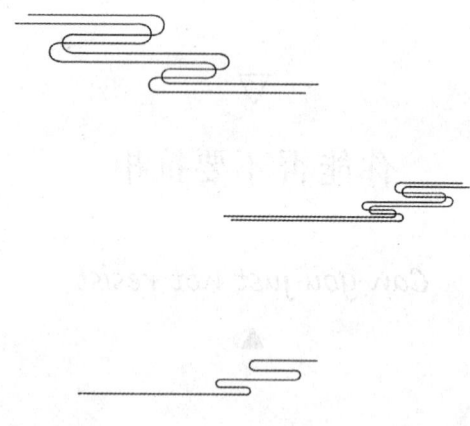

神奇的答案之书
THE BOOK OF MAGIC ANSWERS

▽

放弃之前的打算

Give up the previous plan
▲

▽

有障碍需要克服

There are barriers to be overcome

▲

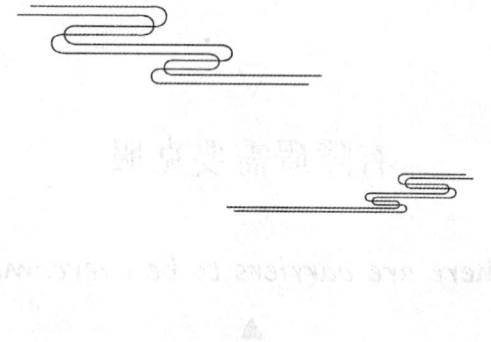

▽

做出改变

Make changes

▲

神奇的答案之书
THE BOOK OF MAGIC ANSWERS

▽

把这看作一次机会

Regard it as an opportunity

▲

▽

别再犹豫了

No more hesitation

▲

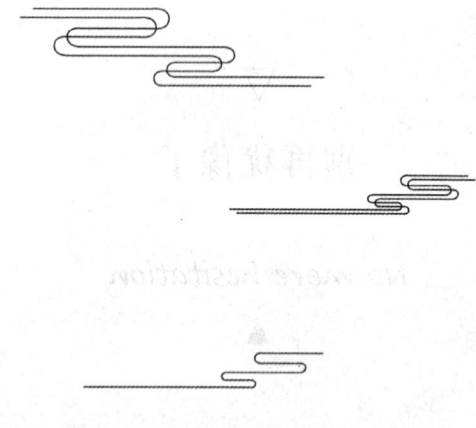

神奇的答案之书
THE BOOK OF MAGIC ANSWERS

▽

保守你的秘密

Keep your secret

▲

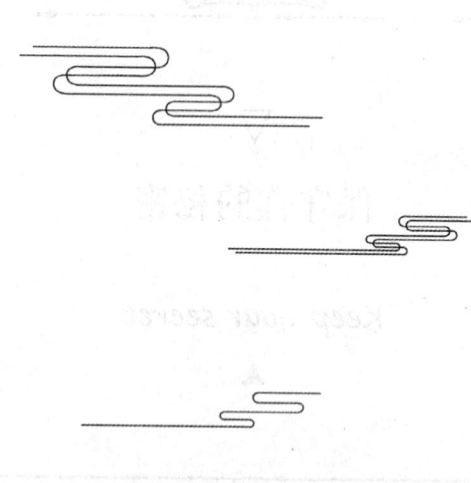

神奇的答案之书
THE BOOK OF MAGIC ANSWERS

▽

采取冒险的态度

Take a risky attitude

▲

神奇的答案之书
THE BOOK OF MAGIC ANSWERS

▽

遵从其他人的建议

Abide by others' suggestions

▲

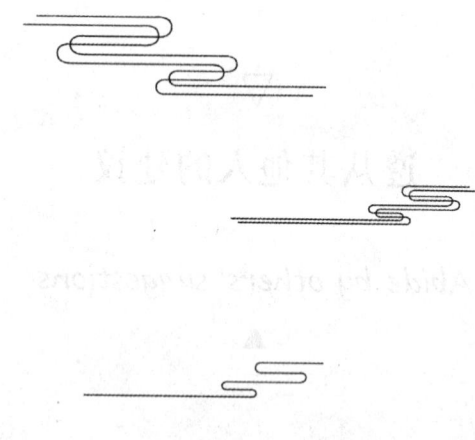

神奇的答案之书
THE BOOK OF MAGIC ANSWERS

▽

改变将不会很快发生

The change will not happen soon

▲

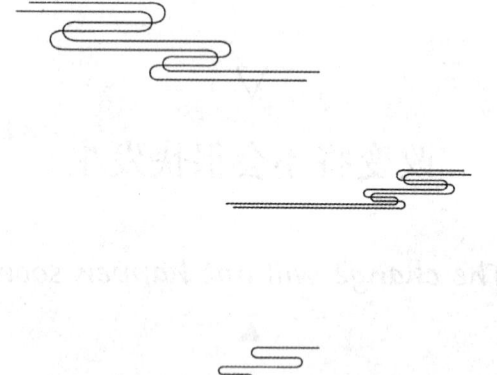

神奇的答案之书
THE BOOK OF MAGIC ANSWERS

▽

似乎已成事实

It seems to have become a fact

▲

负起责任来

Take your responsibility

神奇的答案之书
THE BOOK OF MAGIC ANSWERS

你终会搞清楚你想知道的一切

You will finally make clear everything you want to know

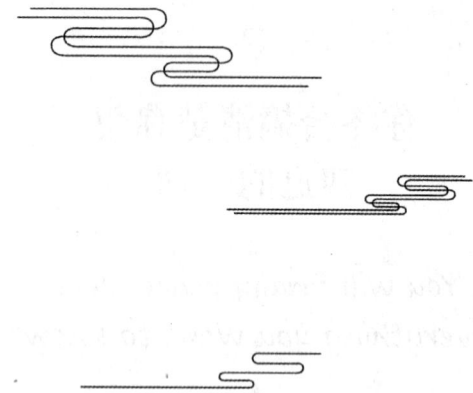

神奇的答案之书
THE BOOK OF MAGIC ANSWERS

▽

且行且思

Think while you act

▲

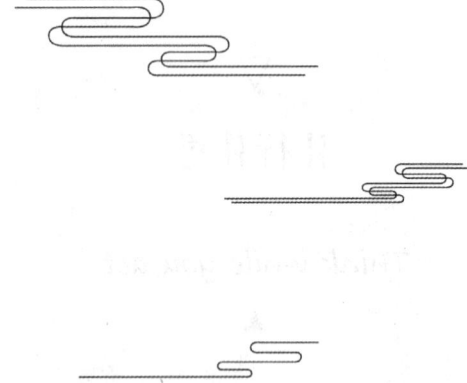

神奇的答案之书
THE BOOK OF MAGIC ANSWERS

▽

如果做就做好，
否则就不要做

Don't do it or do it best

▲

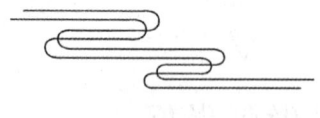

神奇的答案之书
THE BOOK OF MAGIC ANSWERS

▽

不值得一争

It is not worth a dispute

▲

不值得一提

It is not worth a mention

神奇的答案之书
THE BOOK OF MAGIC ANSWERS

▽

你肯定会获得支持

You will certainly have supporters

▲

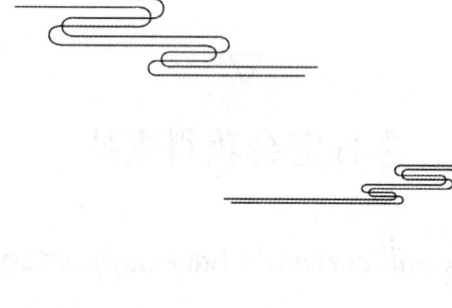

神奇的答案之书
THE BOOK OF MAGIC ANSWERS

神奇的答案之书
THE BOOK OF MAGIC ANSWERS

▽

向别人倾诉

Pour out to others

▲

▽

绝对不会

Absolutely not

▲

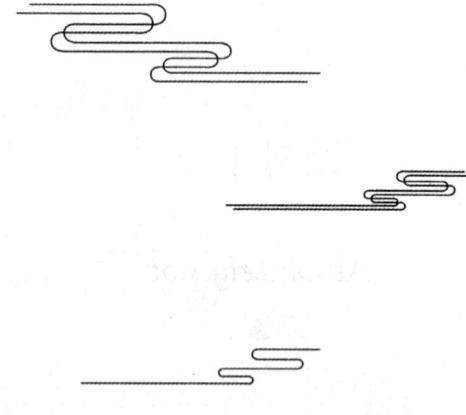

神奇的答案之书
THE BOOK OF MAGIC ANSWERS

神奇的答案之书
THE BOOK OF MAGIC ANSWERS

▽

等待一个更好的提议

Wait for a better suggestion

▲

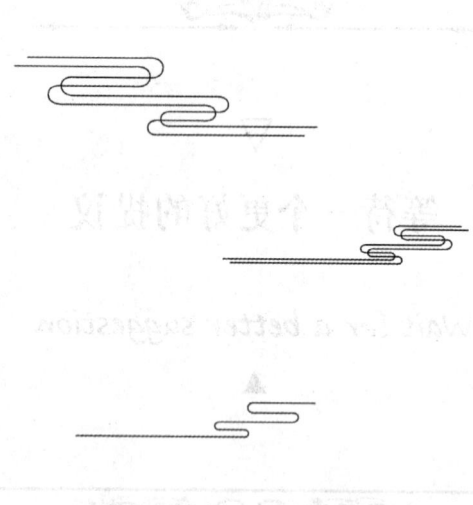

▽

可能会惹上麻烦

You may get into trouble

▲

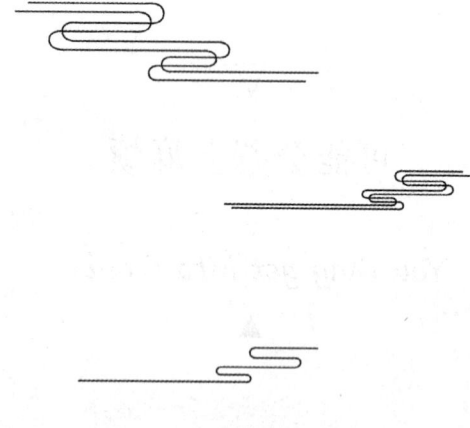

神奇的答案之书
THE BOOK OF MAGIC ANSWERS

神奇的答案之书
THE BOOK OF MAGIC ANSWERS

▽

你有可能会遭遇反对

It's possible for you to meet with opposition

▲

敞开的窗子理想以及

It is possible for you to meet with approval.

神奇的答案之书
THE BOOK OF MAGIC ANSWERS

告诉某人那对你意味着什么

Tell somebody what that means to you

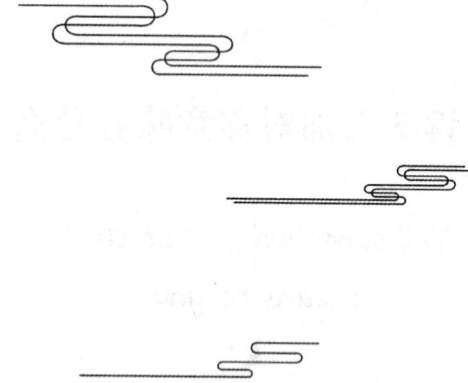

神奇的答案之书
THE BOOK OF MAGIC ANSWERS

▽

把这看作一个时机

Regard it as an opportunity

▲

▽

不要陷入坏情绪

Don't fall into a bad mood

▲

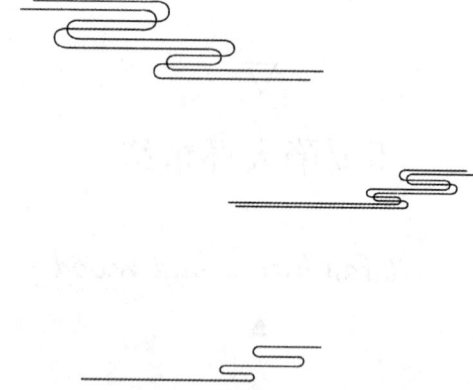

▽

享受这个过程

Enjoy this process

▲

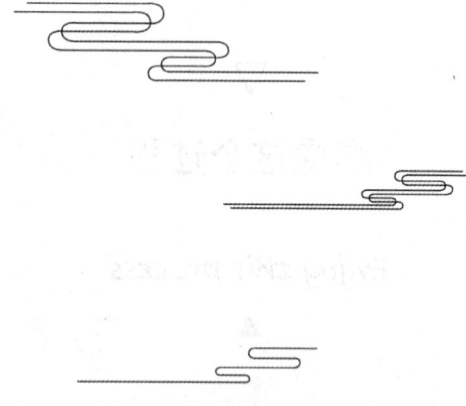

神奇的答案之书
THE BOOK OF MAGIC ANSWERS

▽

不要勉强自己

Don't force yourself

▲

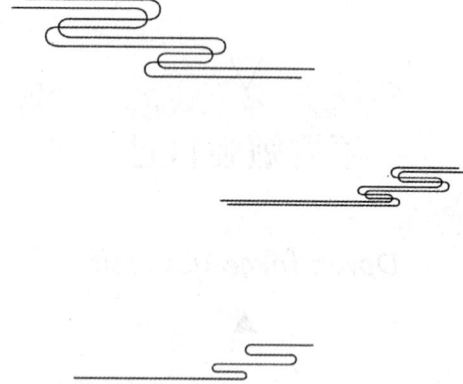

神奇的答案之书
THE BOOK OF MAGIC ANSWERS

▽

你不得不放弃其他一些东西

You have to give up some other things

▲

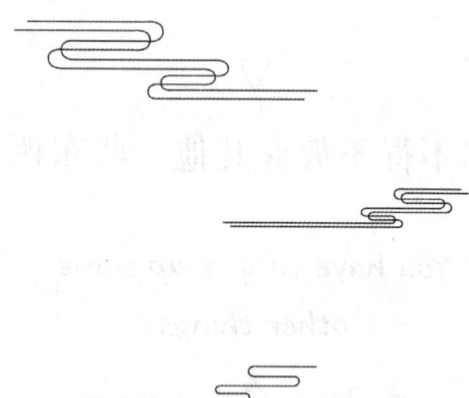

神奇的答案之书
THE BOOK OF MAGIC ANSWERS

神奇的答案之书
THE BOOK OF MAGIC ANSWERS

▽

先做好自己的事

Get your own things done first

▲

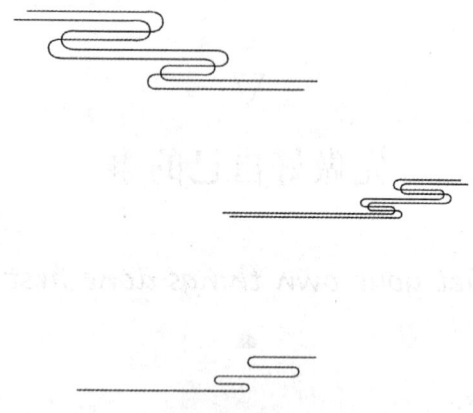

神奇的答案之书
THE BOOK OF MAGIC ANSWERS

▽

享受这种经历

Enjoy the experience

▲

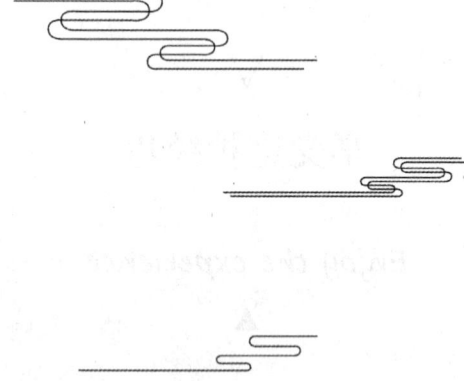

神奇的答案之书
THE BOOK OF MAGIC ANSWERS

神奇的答案之书
THE BOOK OF MAGIC ANSWERS

▽

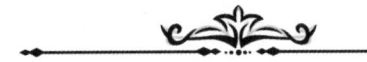

你需要考虑其他方法

You need to consider
other methods

▲

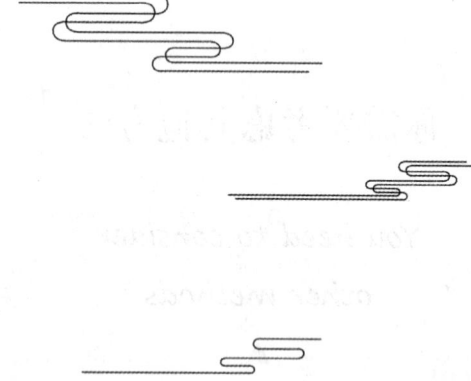

神奇的答案之书
THE BOOK OF MAGIC ANSWERS

▽

值得付出努力

It is worth your effort

▲

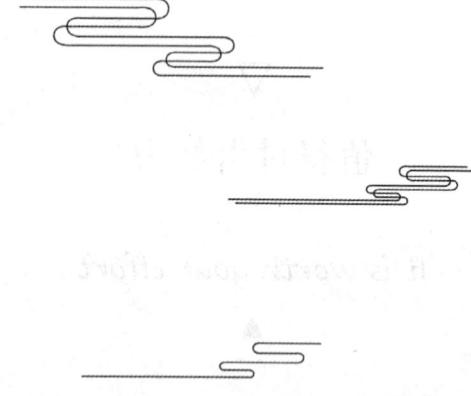

▽

你最好等一等

You'd better wait

▲

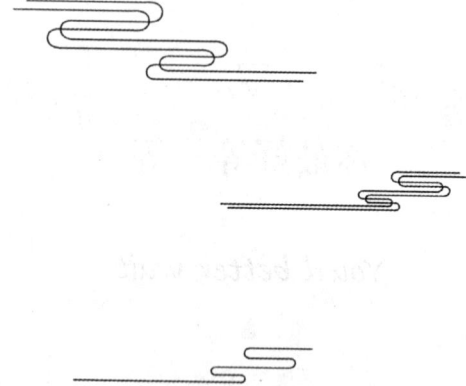

神奇的答案之书
THE BOOK OF MAGIC ANSWERS

神奇的答案之书
THE BOOK OF MAGIC ANSWERS

▽

欣然接受

Accept with pleasure

▲

○●○●○○●○●●●●···

神奇的答案之书
THE BOOK OF MAGIC ANSWERS

神奇的答案之书
THE BOOK OF MAGIC ANSWERS

▽

听从其他人的意见

Obey others' suggestions

▲

神奇的答案之书
THE BOOK OF MAGIC ANSWERS

▽

别那么荒唐

Don't be ridiculous

▲

神奇的答案之书
THE BOOK OF MAGIC ANSWERS

▽

相信你的直觉

Trust your instincts

▲

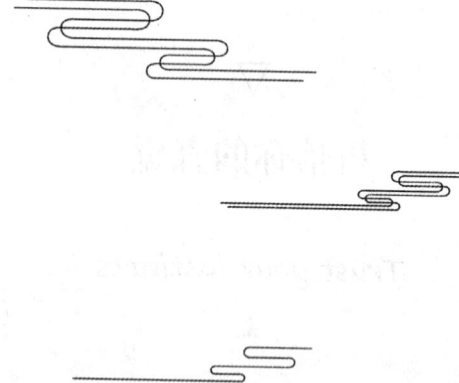

神奇的答案之书
THE BOOK OF MAGIC ANSWERS

▽

一切都取决于你的选择

It all depends upon your option
Obey regulations

▲

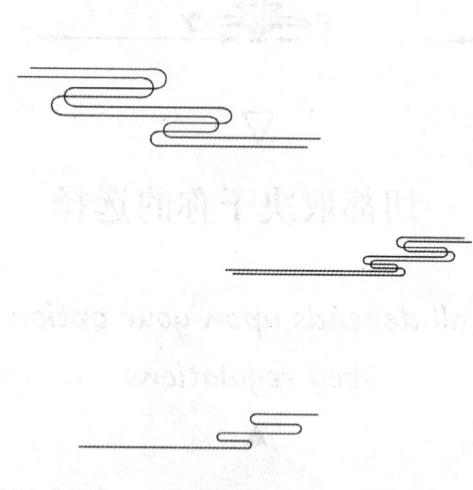

当局者迷

Those closely involved cannot
see clearly

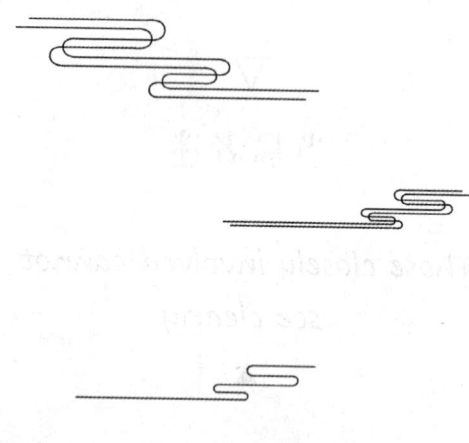

神奇的答案之书
THE BOOK OF MAGIC ANSWERS

最好的解决方法还不明朗

The best method is not clearly known

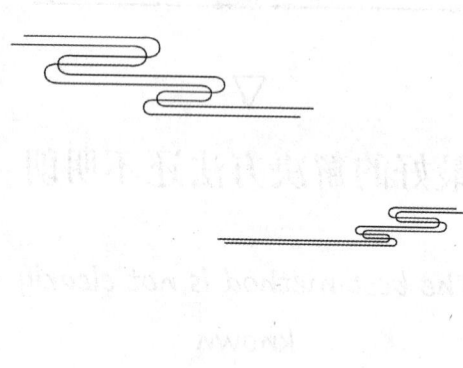

▽

明天再试试

Try again tomorrow

▲

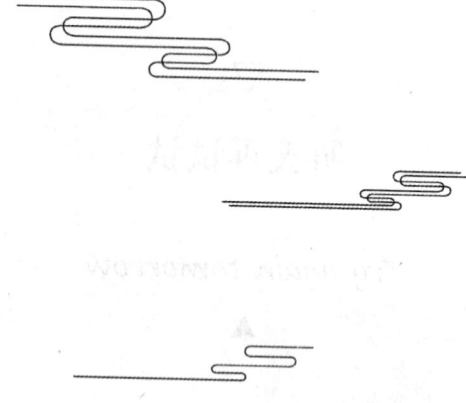

神奇的答案之书
THE BOOK OF MAGIC ANSWERS

▽

遵守规定

Obey regulations

▲

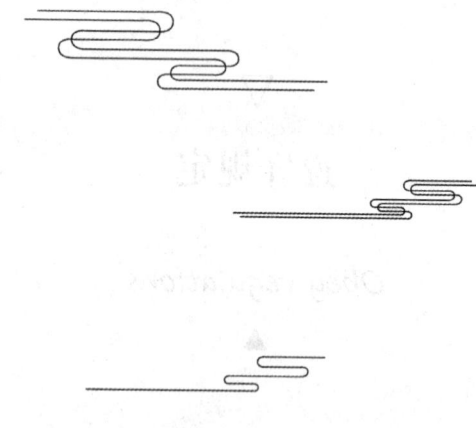

神奇的答案之书
THE BOOK OF MAGIC ANSWERS

▽

不识庐山真面目，
只缘身在此山中。

It can not be seen truthfully, just because you are inside it

▲

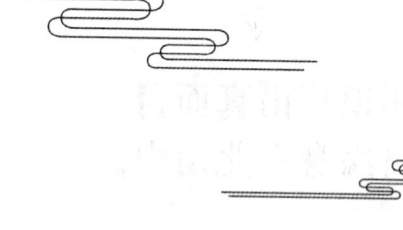

神奇的答案之书
THE BOOK OF MAGIC ANSWERS